BestMasters

Springer awards „BestMasters" to the best master's theses which have been completed at renowned universities in Germany, Austria, and Switzerland.

The studies received highest marks and were recommended for publication by supervisors. They address current issues from various fields of research in natural sciences, psychology, technology, and economics.

The series addresses practitioners as well as scientists and, in particular, offers guidance for early stage researchers.

Sven Painer

Variation Based Dense 3D Reconstruction

Application on Monocular Mini-Laparoscopic Sequences

With a preface by Prof. Dr.-Ing. Rolf-Rainer Grigat

 Springer Vieweg

Sven Painer
Hamburg, Germany

Awarded with the Fokusfinder Preis 2015 of Initiative Bilderverarbeitung e.V. and the Karl H. Ditze Preis 2015 of Hamburg University of Technology.

OnlinePlus material to this book can be available on http://www.springer-vieweg.de/978-3-658-12697-1

BestMasters
ISBN 978-3-658-12697-1   ISBN 978-3-658-12698-8 (eBook)
DOI 10.1007/978-3-658-12698-8

Library of Congress Control Number: 2016934715

Springer Vieweg

Printed on acid-free paper

This Springer Vieweg imprint is published by Springer Nature
The registered company is Springer Fachmedien Wiesbaden GmbH

Institute Profile

The resarch focus of TUHH Vision Systems, a research group of Hamburg University of Technology, is on image analysis and pattern recognition, both in basic research and its transfer to industrial applications.

Examples of our research expertise are three-dimensional geometric reconstruction from video sequences (3D Computer Vision), color management and multispectral modeling, camera calibration and camera characterization, medical image analysis, and methods for classification, detection and parameter estimation. Methods are related to mathematics, physics, communications engineering, microelectronics and computer science.

Selected application fields are 3D modeling of surfaces, self-calibration of cameras, reconstruction of indoor scenes featuring only sparse or no texturing, inverse problems in medicine, joint segmentation in x-ray images, optimal color reproduction, and low-level algorithms for mobile phone cameras.

Preface

3D modeling of cirrhotic liver surfaces based on laparoscopic videos will render possible the quantification of fibrosis in almost real-time. Up to now, qualitative and verbal descriptions as well as video documentations had to be used.

Origin was the research of Dr.-Ing. Jan Marek Marcinczak, who has been supervising and supporting Mr. Painer's Master Thesis. During his PhD research Mr. Marcinczak has estimated the camera motion and the geometry of the organ surface based on complete video sequences [13]. His methods work off-line and non-causal. Based on this work, Mr. Painer has developed a causal method, which basically works on-line and in real-time.

Mr. Painer parallelized the algorithms and proved the performance of his implementation. He adjusted the methods for the complete reconstruction, consisting of a sparse reconstruction via Structure from Motion and a dense reconstruction via a variational approach, and embedded these into a software framework designed entirely by himself.

The computationally intensive dense reconstruction has been parallelized by Mr. Painer using CUDA and a self-designed class library to be executed on the graphics card. This class library takes care of clearing resources after usage to prevent resource leaks in the context of CUDA. The implementation takes only 55 seconds for the reconstruction of 30 frames consisting of sparse and dense reconstruction. Previous, non-causal algorithms required 30 minutes for the reconstruction.

Mr. Painer's Master Thesis has been awarded the Fokusfinder Preis 2015 of Initiative Bilderverarbeitung e.V. and the Karl H. Ditze Preis 2015 of Hamburg University of Technology.

The topic was motivated by Prof. Dr. Ansgar W. Lohse, Medical Director of I. Department of Internal Medicine of Universitätsklinikum Hamburg-Eppendorf (UKE). The cooperation has been accompanied by PD Dr. Ulrike Denzer, Dr. Julian Holzhüter and Dr. Tobias Werner of UKE. We cordially appreciate their cooperation.

Mr. Painer has skillfully solved all mathematically sophisticated difficulties during his Master Thesis and provided a scalable concurrent implementation, which can basically be executed in real-time on an adequate hardware platform. The result paves the way for supporting the clinicians both on-line during video capturing and for quantitative documentation. The feedback during the intervention can help improving the quality of the videos and the subsequent reconstructions. I hope for a positive response and a broad audience of this work.

Prof. Dr.-Ing. Rolf-Rainer Grigat
TUHH Vision Systems
Hamburg University of Technology

Acknowledgements

I would like to thank my supervisor Dr.-Ing. Jan Marek Marcinczak for guiding me through my work. I would not have come so far without your friendly and competent help. You will be an example for me on how to supervise people.

I would also like to thank Prof. Dr.-Ing. Rolf-Rainer Grigat for giving me the chance of writing this thesis and for all the experiences I was able to gain throughout the time.

A special thank goes to all people of the TUHH Vision Systems institute for the nice working atmosphere. It was a pleasure spending the time with you.

If there are only a few mistakes in this thesis, it is because of the great support by Pia Beilfuß and Michael Garben. Thank you for proofreading this thesis.

Last but not least, I would also like to thank my family for the great support. Without you I would not be where I am today and without you I would not have written this thesis. Thank you very much!

Contents

List of Figures

List of Tables

Chapter 1

Introduction

Mini-laparoscopy is a method to visually inspect the abdomen. It can be used to examine the surface of the liver and detect fibrosis or cirrhosis. At the moment, it is not possible to get any quantitative measurements. Therefore, it is desirable to reconstruct the liver surface and to be able to do quantitative measurements on this reconstruction. In this thesis, a method to reconstruct the liver surface from monocular mini-laparoscopic sequences shall be implemented and evaluated.

This thesis consists of five chapters. In Chapter 1, a motivation to the topic and a task definition is presented. An introduction to the theoretical background is given in Chapter 2. Chapter 3 describes some aspects of the implementation and an evaluation is given in Chapter 4. Finally, a conclusion of the results as well as an outlook for future work is contained in Chapter 5.

1.1 Motivation

Liver cirrhosis is a severe disease. In the U.S., chronic liver diseases and cirrhosis have been the 12th most cause of death in 2011 with an increasing number [9]. The World Health Organization evaluated liver cirrhosis as the 10th most cause of death for men in 2012 [26]. In order to help concerned people, it is important to recognize the liver disease in an early stage. There are different examinations to find liver diseases. Standard procedures include biopsy and fibroscan but these are not as exact as the mini-laparoscopy [10].

During mini-laparoscopy, a thin laparoscope with a diameter of 1.9 mm is used to inspect the abdomen. This procedure is minimal-invasive and the patient can be released the same or the next day from the clinic. As the clinician can see the liver surface, he is able to do a more exact diagnosis than by only having some measurements like the stiffness of the liver. Figure 1.1 shows two examples of images taken from mini-laparoscopic sequences. Figure 1.1 (a) shows a scarred liver surface and Figure 1.1 (b) shows a liver surface with severe cirrhosis.

Mini-laparoscopy has the drawback of not giving quantitative measurements. At the moment, the clinical report contains only a macroscopic description of the liver surface. If the examination is repeated after some time to detect changes, it is hard for another clinician to realize the amount of changes of the liver surface. One possibility to solve this problem is to reconstruct the liver surface and to do quantitative measurements on this reconstruction.

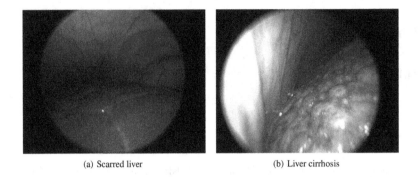

(a) Scarred liver   (b) Liver cirrhosis

Figure 1.1: Two examples of mini-laparoscopic images. (a) shows a scarred liver surface and (b) shows a liver surface with severe cirrhosis. These images are taken from real mini-laparoscopic sequences by courtesy of Universitätsklinikum Hamburg-Eppendorf (UKE).

1.2 Task Definition

Marcinczak et al. have proposed a method to do a dense reconstruction of the liver surface for one reference frame of a monocular mini-laparoscopic sequence [14]. They use a variation based approach for the reconstruction. This approach needs many images of the same region from different viewports in order to get information about the scene. The computations are done offline. That means that recording of the sequence and the reconstruction are done at different times and not in parallel. On the one hand this has the advantage that they are able to inspect the whole sequence to find the best frames for the reconstruction. On the other hand this has the disadvantage that regions with too few frames to get information for the reconstruction will be detected after the recording. At that time, the intervention has been finished and it is not possible to get additional frames for these regions. It is desirable to be able to do the reconstruction online in order to be able to assist the clinicians during recording and show them the regions with too few information. The current implementation is too slow for this application.

A new implementation of all reconstruction steps has to be created that is able to do the reconstruction online. Most of the steps of the dense reconstruction can be done in parallel for each pixel individually. In the last years, the computing power of the graphics cards has been made available for general calculations. These graphics cards are able to compute small but highly parallel tasks very efficiently. Therefore, the implementation should try to use this computing power to do the dense reconstruction. In the end, it shall be evaluated whether it will be possible to do the reconstruction in real-time in the near future. Then it would be possible to assist the clinicians during the recording of the sequences.

Chapter 2

Theoretical Background

This chapter describes the theoretical background of this thesis. It starts with defining the Legendre-Fenchel transformation that is a prerequisite for the dense reconstruction. Then the possibility to be invariant to illumination change by using photometric invariants is introduced. After these prerequisites, the sparse reconstruction will be defined as its results are needed by the dense reconstruction. In the end, the theory of the dense reconstruction is shown.

2.1 Legendre-Fenchel Transformation

It is common to use the dual problem to solve an optimization problem. To get the dual form of a problem, it is transformed into another space in order to make solving the problem easier. The problem is then called primal in the original space and dual in the other space. The Legendre-Fenchel transformation can be used to create a dual problem out of a continuous problem that does not have to be differentiable. The dual form will then be convex and differentiable. This section refers to [6].

The Legendre-Fenchel transformation $f^*(p)$ of a function f is defined as

$$f^*(p) = \sup_{x \in \mathbb{R}} \{px - f(x)\} . \tag{2.1}$$

As the supremum of $px - f(x)$ is searched, the derivate of this function has to be 0. This leads to

$$p = f'(x) . \tag{2.2}$$

A tangent at point $(x, f(x))$ will then be defined as

$$f(x) = px - c . \tag{2.3}$$

This can be transformed to

$$c = px - f(x) , \tag{2.4}$$

with c being the negative intersection of the tangent with the y-axis. Comparing (2.1) and (2.4) shows that the Legendre-Fenchel transformation $f^*(p)$ is the negative intersection of a tangent with the slope p and the y-axis. The Lengendre-Fenchel transformation can also be written in a vector notation as

$$f^*(\mathbf{p}) = \sup_{\mathbf{x} \in \mathbb{R}^n} \{\mathbf{x}^\mathsf{T}\mathbf{p} - f(\mathbf{x})\} . \tag{2.5}$$

In this thesis, the Legendre-Fenchel transformation of the Huber norm is needed. The Huber norm is defined as

$$\|x\|_{\varepsilon} = \begin{cases} \frac{\|x\|_2^2}{2\varepsilon} & \text{if } \|x\|_2 \leq \varepsilon \\ \|x\|_1 - \frac{\varepsilon}{2} & \text{otherwise} \end{cases}. \tag{2.6}$$

To compute the Legendre-Fenchel transformation of the Huber norm, both cases have to be calculated separately. The first case is called f_1 and the second case is called f_2.

First, f_1 has to be inserted into (2.5) leading to

$$f_1^*(x) = \sup_{x \in \mathbb{R}^n} \{x^T p - \frac{\|x\|_2^2}{2\varepsilon}\}. \tag{2.7}$$

The derivative has to be calculated and then the equation has to be rearranged to solve for x which is equivalent to $f_1'^{-1}(p)$.

$$f_1'^{-1}(p) = x = \varepsilon p \tag{2.8}$$

The tangent at point $(x, f_1(x))$ is then defined as

$$f_1(x) = x^T p - c. \tag{2.9}$$

By replacing x with $f_1'^{-1}(p)$, the equation becomes

$$f_1(f_1'^{-1}(p)) = f_1'^{-1}(p)^T p - c. \tag{2.10}$$

(2.8) is inserted into (2.10). Rearranging the equation to solve for c leads to the Legendre-Fenchel transformation of f_1

$$f_1^*(p) = c = \frac{1}{2}\varepsilon\|p\|_2^2. \tag{2.11}$$

The condition for the first case in (2.6) has to be changed to be dependent on p. Therefore, (2.8) is inserted into the condition leading to $\|p\|_2 \leq 1$.

The Legendre-Fenchel transformation of the L^1 norm, that is used in f_2, is given in [6] as

$$f_{L^1}^*(p) = \begin{cases} 0 & \text{if } \|p\| \leq 1 \\ \infty & \text{otherwise} \end{cases}. \tag{2.12}$$

Taking into account the condition space for f_2, only the second case of (2.12) has to be used.

The complete Legendre-Fenchel transformation of the Huber norm is then given as

$$f^*(p) = \begin{cases} \frac{1}{2}\varepsilon\|p\|_2^2 & \text{if } \|p\|_2 \leq 1 \\ \infty & \text{otherwise} \end{cases}. \tag{2.13}$$

2.2 Photometric Invariants

When examining the abdomen with a mini-laparoscope, the only light source is mounted at the tip of the mini-laparoscope. This fact leads to a varying illumination as the light source moves with the mini-laparoscope. When doing 3D reconstruction, features in one image have to be matched to features in another image. In order to be able to match these features in mini-laparoscopic sequences, it is necessary to deal with the changing illumination.

One possibility is to model the light source and surface reflectance. This results in models with a high amount of parameters. Debevec et al. used a special setup and more than 2000 frames to estimate the reflectance model of a face [3]. This is even more difficult inside the abdomen.

Another possibility is to transform the colors from the illumination-variant RGB color space into an illumination-invariant color space. When doing this, it is not necessary to estimate any parameters. This section refers to [17].

To understand the meaning of photometric invariants, first the dichromatic reflection model has to be introduced. A color at point $\mathbf{x} = (x,y)^T$ is defined as

$$\mathbf{c}(\mathbf{x}) = (R(\mathbf{x}), G(\mathbf{x}), B(\mathbf{x}))^T . \tag{2.14}$$

It can also be defined as a combination of the interface reflection component $\mathbf{c}_i$ and body reflection component $\mathbf{c}_b$

$$\mathbf{c}(\mathbf{x}) = \mathbf{c}_i(\mathbf{x}) + \mathbf{c}_b(\mathbf{x}) . \tag{2.15}$$

The interface reflection component describes specularities or highlights. The body reflection component describes the reflectance of the matte body.

If spectrally uniform illumination is assumed, (2.15) can be further decomposed into

$$\mathbf{c}(\mathbf{x}) = e(m_i(\mathbf{x})\widehat{\mathbf{c}_i}(\mathbf{x}) + m_b(\mathbf{x})\widehat{\mathbf{c}_b}(\mathbf{x})) , \tag{2.16}$$

with e being the overall intensity, $m(\mathbf{x})$ being the geometrical reflection factor and $\widehat{\mathbf{c}}(\mathbf{x})$ being the reflectance color. As spectrally uniform illumination is assumed, all three channels of $\widehat{\mathbf{c}_i}(\mathbf{x})$ have to be equal. $\widehat{\mathbf{c}_i}(\mathbf{x})$ will then be called $\mathbf{w}_i(\mathbf{x})$.

If also neutral interface reflection is assumed, $\mathbf{w}_i(\mathbf{x})$ becomes independent of $\mathbf{x}$. Then (2.16) becomes

$$\mathbf{c}(\mathbf{x}) = e(m_i(\mathbf{x})w_i\mathbf{1} + m_b(\mathbf{x})\widehat{\mathbf{c}_b}(\mathbf{x})) , \tag{2.17}$$

with $\mathbf{1}$ being the vector $(1,1,1)^T$. Using (2.17), photometric invariance can be defined as a color $\mathbf{c}(\mathbf{x})$ being invariant to at least one parameter e, m_b or m_i.

There are three different classes of photometric invariance characterizing the independence on these three parameters. The first one is only independent on e and therefore handles global multiplicative illumination changes. The second class is independent on e and m_b. This is true at least for matte surfaces ($m_i = 0$) and handles shadow and shading. The last class is independent on all three parameters and handles highlights and specular reflections.

There are different strategies to get photometric invariants. Basically, these are normalization techniques, log-derivatives and transformations to other color spaces. In this thesis, the spherical transformation into the $r\phi\theta$ color space is used. It is defined as

$$(R,G,B)^T \mapsto \begin{cases} r = \sqrt{R^2 + G^2 + B^2} \\[2mm] \theta = \arctan\left(\dfrac{G}{R}\right) \\[2mm] \phi = \arcsin\left(\dfrac{\sqrt{R^2 + G^2}}{\sqrt{R^2 + G^2 + B^2}}\right) \end{cases} . \tag{2.18}$$

In (2.18), θ and ϕ are invariant to shadow and shading and r is no photometric invariant.

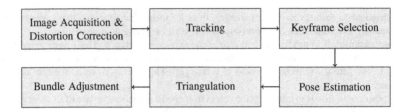

Figure 2.1: Overview of the sparse reconstruction. The necessary steps for the Structure from Motion approach are shown.

2.3 Sparse Reconstruction

When doing 3D reconstruction from a sequence of images, the scene and the camera positions have to be estimated. It is also possible to estimate the intrinsic parameters of the camera, but in this thesis calibrated cameras are used. This has the advantage of becoming more robust because less parameters have to be estimated. A Structure from Motion approach, based on the one proposed by Pollefeys et al. [21], is used to do the sparse reconstruction. It is called sparse reconstruction as it does not give 3D information for each pixel. It works with feature points that are good to track and only gives 3D information for these points.

Figure 2.1 shows the different steps of the Structure from Motion approach that is used. During the first step, the images are captured. Afterwards, a distortion correction is done. Within the second step the points, which are good for tracking, are located. Now these points are matched in different frames. After matching, a keyframe selection has to be done to decide which frames should be used for pose estimation and triangulation. During pose estimation, the positions of the cameras of the different frames will be estimated. The feature points are backprojected into the 3D space during the triangulation step and finally bundle adjustment is done to optimize the positions of the cameras and the 3D points.

2.3.1 Tracking

For reconstructing the camera positions and 3D points, it is necessary to estimate the motion of the image points between two frames first. Therefore, distinctive points have to be found in the frames. These points are called keypoints. From these keypoints, vectors can be extracted that describe the keypoints. These vectors are called descriptors. The descriptors can be matched together to receive the motion of the pixels. This motion can be used to estimate the motion of the cameras later. The Scale Invariant Feature Transform (SIFT) is used to get the keypoints and the descriptors and was introduced by Lowe [12]. SIFT features are invariant to scale and rotation. In addition they are robust to changes in illumination and changes in the viewpoint.

To get robust results of the following steps, long trajectories of matching keypoints over the sequence are important. It is possible that keypoints are outside the viewpoint for a few frames or that single uninformative frames during the sequence appear. These circumstances can become a problem if the matching is only be done between consecutive frames. The trajectories will then end at these frames. Therefore, matching is not only done with the directly preceding frame, but also with some more preceding frames. This could lead to the problem that more than one point has to be inserted into the same trajectory. Schlüter et al. described four strategies to deal with that problem [22]. The strategy *New Trajectory for All* is used in this thesis as Schlüter et al. have also shown that this is the best strategy. It gives the highest absolute number of trajectories and also the highest relative number of correct trajectories. In case of a conflict, none of the conflicting keypoints are inserted into the trajectory and new trajectories are created for all of them.

2.3.2 Keyframe Selection and Pose Estimation

The correspondences resulting from the tracking are noisy. If the motion of the keypoints is very small between two frames, the assumed motion due to noise can be bigger than the real motion of the keypoints. Therefore, the motion of the keypoints between two frames has to be as big as possible to increase the signal-to-noise ratio and to enable a robust estimation of the camera positions.

In addition to the distance of the keypoints, it is also important to have enough correspondences. Normally, five to eight correspondences are sufficient, depending on the algorithm that is chosen for pose estimation, but besides noise rigidity is another problem. These algorithms perform pose estimation for rigid transformations. The liver in case of cirrhosis is rigid, but the deformation of the abdomen during respiration is non-rigid. To handle noise and rigidity, a Random Sample Consensus (RANSAC) approach, as first described by Fischler et al. [4], is used. During RANSAC, a set of points is chosen to be inliers and a corresponding model is estimated. Then all points are checked whether they can be considered being inliers to this model. All inliers are then called the consensus set. The model is taken as a candidate if the number of points in the consensus set is big enough. These steps are repeated a fixed number of times. Models with a greater number of points in the consensus set are taken as the new candidate. In the end, the model with the biggest number of points in the consensus set is taken as the true model and outliers of this model can be discarded. In order to use this RANSAC approach, more correspondences than the minimum number for the algorithm are required. Therefore, besides the condition of the minimum distance of the keypoints, there is also a condition of a minimum number of correspondences.

In this thesis, a simple algorithm for keyframe selection is chosen. Each new frame is compared to the last keyframe. If the motion of the keypoints between these two frames and the number of correspondences are both higher than a threshold, the frame is considered as being a keyframe. Estimation of the camera poses as well as triangulation is only done using the chosen keyframes.

The estimation of the motion between the keyframes is done by using the five-point-algorithm as described by Nistér [19]. This algorithm estimates an essential matrix E between the two frames that is decomposed into a rotation matrix R and a translation vector **t** as also shown in Nistér's paper. First, the matrix

$$D = \begin{pmatrix} 0 & 1 & 0 \\ -1 & 0 & 0 \\ 0 & 0 & 1 \end{pmatrix} \quad (2.19)$$

is defined. Then the essential matrix E is decomposed via singular value decomposition into $\mathrm{Udiag}(1,1,0)\mathrm{V}^{\mathrm{T}}$. U and V are chosen in a way that their determinant is greater than zero. This decomposition is defined up to scale. The translation vector is

$$\mathbf{t} \sim [u_{13}\ u_{23}\ u_{33}]^{\mathrm{T}}\ . \tag{2.20}$$

For the rotation matrix there are the two possible solutions, namely

$$R_a = \mathrm{UDV}^{\mathrm{T}} \tag{2.21}$$

and

$$R_b = \mathrm{UD}^{\mathrm{T}}\mathrm{V}^{\mathrm{T}}\ . \tag{2.22}$$

The four possible transformation matrices are $P_A = [R_a|\mathbf{t}]$, $P_B = [R_a|-\mathbf{t}]$, $P_C = [R_b|\mathbf{t}]$ and $P_D = [R_b|-\mathbf{t}]$, but only one is physically valid. In this valid configuration, the points are in front of both cameras. In the other three solutions, the points are behind one or both of the cameras.

There can be problems with motions that are nearly sidewards as von Öhsen et al. have shown [24]. They have also proposed an algorithm to handle these problems. The first step is to check whether the motion is sidewards. This is done by evaluating the inequality

$$\|\mathrm{I} - \mathrm{R}\|_F \le 0.05\ , \tag{2.23}$$

with $\|\cdot\|_F$ being the Frobenius norm and I being the identity matrix. If this inequality is fulfilled, the correct direction $\tilde{\mathbf{t}}$ of the translation vector is determined as

$$\tilde{\mathbf{t}} = \begin{cases} \mathbf{t} & \text{if median}(\mathbf{x} - \mathbf{x}') > 0 \\ -\mathbf{t} & \text{if median}(\mathbf{x} - \mathbf{x}') \le 0 \end{cases} . \tag{2.24}$$

2.3.3 Triangulation

The next step after estimating the camera positions is to reconstruct the 3D points. When the camera parameters are known, it is possible to exactly calculate the point in the image plane $\mathbf{x}$ where a 3D point $\mathbf{X}$ is being projected to by calculating

$$\mathbf{x} = \mathrm{KPX}\ . \tag{2.25}$$

$K \in \mathbb{R}^{3 \times 3}$ contains the intrinsic camera parameters and is defined as

$$K = \begin{pmatrix} \alpha_x & s & x_0 \\ & \alpha_y & y_0 \\ & & 1 \end{pmatrix}\ , \tag{2.26}$$

with $\alpha_x = f m_x$ and $\alpha_y = f m_y$ being the focal lengths in pixel dimensions in x- and y-direction, $(x_0, y_0)^{\mathrm{T}}$ being the principal point in pixel dimensions and s being the skew factor. $P \in \mathbb{R}^{3 \times 4}$ contains the extrinsic camera parameters and is defined as

$$P = [R|\mathbf{t}]\ , \tag{2.27}$$

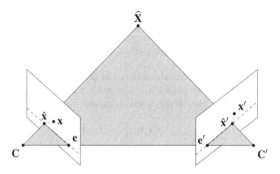

Figure 2.2: Overview of the epipolar geometry. The camera centers are located at $\mathbf{C}$ and $\mathbf{C}'$ and the epipoles are denoted by $\mathbf{e}$ and $\mathbf{e}'$. The 3D point $\widehat{\mathbf{X}}$ is projected into both images onto the points $\hat{\mathbf{x}}$ and $\hat{\mathbf{x}}'$, respectively. These could be shifted to $\mathbf{x}$ and $\mathbf{x}'$ due to noise. In both image planes the epipolar lines are given by a dashed line.

with $R \in \mathbb{R}^{3 \times 3}$ being the rotation matrix and $\mathbf{t} \in \mathbb{R}^3$ being the translation vector. Details about the camera parameters can be found in [7].

It can also be seen by the matrix dimensions that the opposite problem, to find a 3D point corresponding to an image point, can not be solved unambiguously. It is only possible to determine a line on which the point has to be. Therefore, a corresponding point in another image has to be known. Figure 2.2 illustrates this situation. If the two cameras and the image points $\hat{\mathbf{x}}$ and $\hat{\mathbf{x}}'$ are known, it is possible to find the 3D point $\widehat{\mathbf{X}}$ by constructing the lines through the camera centers and the image points, and determining the intersection of these two lines. This is called triangulation.

The problem is that the correspondences are noisy and therefore the lines do not intersect. In order to have an intersection, the epipolar constraint

$$\hat{\mathbf{x}}'^{T} F \hat{\mathbf{x}} = 0 \tag{2.28}$$

has to be fulfilled. F is the fundamental matrix and is related to the essential matrix E by

$$E = K'^{T} F K \,. \tag{2.29}$$

The case of noisy correspondences is illustrated by the points $\mathbf{x}$ and $\mathbf{x}'$ in Figure 2.2. These points do not fulfill (2.28). Hartley and Zisserman proposed a method to correct these correspondences in order to fulfill the epipolar constraint and to enable a simple triangulation method [7]. They minimize the function

$$C(\mathbf{x}, \mathbf{x}') = d(\mathbf{x}, \hat{\mathbf{x}})^2 + d(\mathbf{x}', \hat{\mathbf{x}}')^2 \tag{2.30}$$

with respect to the epipolar constraint. The function $d(\cdot, \cdot)$ denotes the euclidean distance between the two points. A pair of points, that fulfills the epipolar constraint, has to lie on two corresponding epipolar lines $\mathbf{l}$ and $\mathbf{l}'$. In Figure 2.2, the epipolar lines are given by dashed lines.

All points on these lines will fulfill the epipolar constraint, but only the ones being nearest to $\mathbf{x}$ and $\mathbf{x}'$ will minimize the function $C(\mathbf{x}, \mathbf{x}')$. Therefore, (2.30) can be written as

$$C(\mathbf{x}, \mathbf{x}') = d(\mathbf{x}, \mathbf{l})^2 + d(\mathbf{x}', \mathbf{l}')^2 , \tag{2.31}$$

where $d(\mathbf{x}, \mathbf{l})$ now denotes the perpendicular distance of the point $\mathbf{x}$ to the line $\mathbf{l}$. The lines $\mathbf{l}$ and $\mathbf{l}'$ can be all possible pairs of epipolar lines. The points $\hat{\mathbf{x}}$ and $\hat{\mathbf{x}}'$ are then chosen to be the closest points to $\mathbf{x}$ and $\mathbf{x}'$ on the lines. If the pencil of epipolar lines is parametrized by t, (2.31) becomes

$$\min_t C(\mathbf{x}, \mathbf{x}') = d(\mathbf{x}, \mathbf{l}(t))^2 + d(\mathbf{x}', \mathbf{l}'(t))^2 . \tag{2.32}$$

Details on the minimization can be found in [7].

2.3.4 Bundle Adjustment

After estimating the camera positions and 3D points, an optimization step has to be done. For each 3D point, there are many corresponding 2D points in different frames because of the trajectories found during tracking. Normally, for each 2D point

$$\mathbf{x}_j^i = \mathrm{P}^i \mathbf{X}_j \tag{2.33}$$

has to be fulfilled, but due to noise this is not the case. Therefore, projection matrices $\hat{\mathrm{P}}^i$ and 3D points $\hat{\mathbf{X}}_j$ that fulfill (2.33) and minimize the geometric error between the reprojected image point and the measured image point have to be found. This leads to

$$\min_{\hat{\mathrm{P}}^i, \hat{\mathbf{X}}_j} \sum_i \sum_j d(\hat{\mathrm{P}}^i \hat{\mathbf{X}}_j, \mathbf{x}_j^i)^2 , \tag{2.34}$$

where $d(\cdot, \cdot)$ denotes the geometric image distance between the two points. This minimization is called bundle adjustment and can be found in [7].

2.4 Dense Reconstruction

The sparse reconstruction gives 3D points for features that are suitable to track. If there are no features in a region of interest, there will also be no 3D information about this region. The dense reconstruction, in contrast to the sparse reconstruction, gives depth information for each pixel of a reference frame and therefore 3D information for each pixel. The dense reconstruction is done by using a variational approach and is handled in detail in this section. This section refers to [18].

2.4.1 Variational Approach

In a variational approach, the solution of a problem is considered as an energy functional and this energy functional has to be minimized. Horn and Schunck are considered to be the first to use a variational approach in computer vision [8]. They applied it to compute the optical flow between two consecutive images in a sequence. Wedel and Cremers refined this approach in [25]. Newcombe et al. used a variational approach to do a dense reconstruction [18].

Newcombe et al. proposed a method to compute an inverse depth map ξ. It is called inverse because the values of this map are the inverse of the depth. They formulated the energy E_ξ, which has to be minimized to compute the inverse depth map, as

$$E_\xi = \int_\Omega \{g(\mathbf{u})\|\nabla\xi(\mathbf{u})\|_\varepsilon + \lambda C(\mathbf{u},\xi(\mathbf{u}))\}d\mathbf{u} \ . \tag{2.35}$$

This equation consists of a regularizer and a cost volume. The regularizer is given by $\|\nabla\xi(\mathbf{u})\|_\varepsilon$ and the cost volume is given by $C(\mathbf{u},\xi(\mathbf{u}))$. The cost volume contains the data terms and will be defined in detail in Section 2.4.2. It is weighted by the factor λ. In the regularizer, $\|\cdot\|_\varepsilon$ denotes the Huber norm, which is defined in (2.6). The regularizer is weighted for each pixel by the function $g(\mathbf{u})$, which is defined as

$$g(\mathbf{u}) = e^{-\alpha\|\nabla I_r(\mathbf{u})\|_2^\beta} \ , \tag{2.36}$$

where $I_r(\mathbf{u})$ is the intensity of the pixel at location $\mathbf{u}$. This weighting function decreases the influence of the regularizer at edges so that they can remain in the inverse depth map.

2.4.2   Cost Volume

The cost volume $C(\mathbf{u},\xi(\mathbf{u}))$ encapsulates the data terms and therefore gives information about the image content. It is a three-dimensional space, that maps each depth sample d for each pixel $\mathbf{u}$ of a reference frame r to costs for this tuple. The computation is done as

$$C_r(\mathbf{u},d) = \frac{1}{N_{r,\text{inside}}(\mathbf{u},d)} \sum_{m\in\mathcal{I}(r)} \|\rho_r(I_m,\mathbf{u},d)\|_1 \ . \tag{2.37}$$

For each pixel $\mathbf{u}$ and each depth sample d, the backprojected 3D points are projected into all other frames nearby, $\mathcal{I}(r)$, and the L^1 norms of the photometric errors $\rho_r(I_m,\mathbf{u},d)$ are summed up. The number of projections that are inside the other frames is denoted as $N_{r,\text{inside}}(\mathbf{u},d)$ and I denotes the image. Figure 2.3 illustrates the calculation of the cost volume.

For the computation of the cost volume, the photometric error is used that is defined as

$$\rho_r(I_m,\mathbf{u},d) = I_r(\mathbf{u}) - I_m(\pi(KT_{rm}\pi^{-1}(\mathbf{u},d))) \ , \tag{2.38}$$

where T_{rm} denotes the transformation of a point in the coordinate system of camera r into the coordinate system of camera m. The dehomogenisation of a point $\mathbf{x} = (x,y,z)^T$ is given as $\pi(\mathbf{x}) = (x/z,y/z)^T$ and $\pi^{-1}(\mathbf{u},d) = \frac{1}{d}K^{-1}\mathbf{u}$ is the backprojection of an image point into the 3D space. In (2.38), the data terms of the images at a pixel are compared. Newcombe et al. used gray values as data terms but they are not invariant to illumination changes. Marcinczak et al. used spherical data terms in order to cope with illumination changes [14]. Therefore, the $r\phi\theta$ color space is used. The transformation into this color space can be found in (2.18). When applying the spherical transformation, the photmetric error becomes

$$\rho_r(I_m,\mathbf{u},d) = \begin{bmatrix} \theta_r(\mathbf{u}) - \theta_m(\pi(KT_{rm}\pi^{-1}(\mathbf{u},d))) \\ \phi_r(\mathbf{u}) - \phi_m(\pi(KT_{rm}\pi^{-1}(\mathbf{u},d))) \end{bmatrix} \ . \tag{2.39}$$

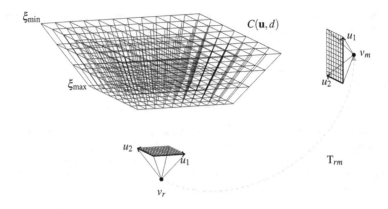

Figure 2.3: Illustration of the procedure to compute the cost volume. All pixels of the reference frame v_r are backprojected into the 3D space for all depth samples d between ξ_{min} and ξ_{max}. In this example, there are three depth samples. This voxel volume is then projected into another frame v_m. The values of the pixel in the reference frame and the pixel in the other frame are then taken to compute the cost volume. This figure is taken from [14].

As the component r of the $r\phi\theta$ color space is no photometric invariant, it is not used in (2.39).

An inverse depth map can be constructed from the cost volume by using the inverse depth value with the minimum costs for each pixel. Figure 2.4 (a) shows such a resulting inverse depth map. It can be seen that this inverse depth map is very noisy. Figure 2.4 (b) shows the photometric error for the pixel $\mathbf{u}$ that is marked in Figure 2.4 (a). Ideally, there should be only one distinct minimum, but in regions with low information this is not the case and leads to the noise in the resulting inverse depth map. Therefore, some regularization has to be done on the inverse depth map as described in Section 2.4.4.

2.4.3 Coupling to Sparse Reconstruction

The dense reconstruction needs information from the sparse reconstruction. On the one hand, the camera positions have to be known. For the construction of the cost volume, the transformation from the reference frame to the frames nearby has to be known. This knowledge is inevitable. When there is no information about the scene so far, this information comes from the sparse reconstruction. Later, it can also come from tracking in the dense domain.

On the other hand, the 3D points from the sparse reconstruction can be used to guide the dense reconstruction. Tests have shown that coupling the dense reconstruction to the result of the sparse reconstruction gives considerably better results. However, there have been no quantitative evaluations. There are two possibilities to couple the dense reconstruction to the

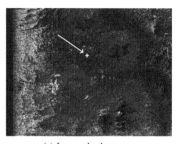
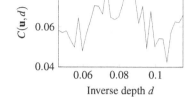

(a) Inverse depth map   (b) Photometric error for the marked pixel **u**

Figure 2.4: Inverse depth map constructed from cost volume. (a) shows an inverse depth map constructed only by the cost volume. For each pixel, the inverse depth value with the minimum costs is taken. (b) shows the error function for the pixel **u** that is marked in (a).

information about the 3D points. In both possibilities, an inverse depth map has to be interpolated from the 3D points. This interpolated inverse depth map will be called $\xi_i(\mathbf{u})$. One possibility is to use $\xi_i(\mathbf{u})$ as the initial depth map when starting the minimization. The other possibility is to change (2.35) to become

$$E_\xi = \int_\Omega \{g(\mathbf{u})\|\nabla\xi(\mathbf{u})\|_\varepsilon + \lambda C(\mathbf{u},\xi(\mathbf{u})) + \lambda_2(\xi(\mathbf{u}) - \xi_i(\mathbf{u}))^2\}d\mathbf{u} . \qquad (2.40)$$

The term $(\xi(\mathbf{u}) - \xi_i(\mathbf{u}))^2$ is added in order to couple the inverse depth map to the results of the sparse reconstruction. This term is weighted by λ_2.

2.4.4 Minimization

In order to minimize the energy given in (2.40), a technique called quadratic relaxation can be used. Details about the quadratic relaxation can be found in [25, 29]. During this step, an auxiliary variable α is introduced, such that (2.40) becomes

$$E_{\xi,\alpha} = \int_\Omega \{g(\mathbf{u})\|\nabla\xi(\mathbf{u})\|_\varepsilon + \frac{1}{2\theta}(\xi(\mathbf{u}) - \alpha(\mathbf{u}))^2$$
$$+ \lambda C(\mathbf{u},\alpha(\mathbf{u})) + \lambda_2(\alpha(\mathbf{u}) - \xi_i(\mathbf{u}))^2\}d\mathbf{u} . \qquad (2.41)$$

The new term $Q(\mathbf{u}) = \frac{1}{2\theta}(\xi(\mathbf{u}) - \alpha(\mathbf{u}))^2$ penalizes differences between $\xi(\mathbf{u})$ and $\alpha(\mathbf{u})$ and ensures that $\xi(\mathbf{u}) = \alpha(\mathbf{u})$ for $\theta \to 0$. In this case, (2.41) is equal to (2.40).

To get the inverse depth map ξ, (2.40) has to be minimized:

$$\min_{\xi,\alpha} \int_\Omega \{g(\mathbf{u})\|\nabla\xi(\mathbf{u})\|_\varepsilon + \frac{1}{2\theta}(\xi(\mathbf{u}) - \alpha(\mathbf{u}))^2$$
$$+ \lambda C(\mathbf{u},\alpha(\mathbf{u})) + \lambda_2(\alpha(\mathbf{u}) - \xi_i(\mathbf{u}))^2\}d\mathbf{u} . \qquad (2.42)$$

Because of the quadratic relaxation, it is possible to split (2.42) into

$$\min_{\xi} \int_{\Omega} \{g(\mathbf{u}) \|\nabla \xi(\mathbf{u})\|_{\varepsilon} + \frac{1}{2\theta}(\xi(\mathbf{u}) - \alpha(\mathbf{u}))^2\} d\mathbf{u} \qquad (2.43)$$

and

$$\min_{\alpha} \int_{\Omega} \{\frac{1}{2\theta}(\xi(\mathbf{u}) - \alpha(\mathbf{u}))^2 + \lambda C(\mathbf{u}, \alpha(\mathbf{u})) + \lambda_2(\alpha(\mathbf{u}) - \xi_i(\mathbf{u}))^2\} d\mathbf{u} . \qquad (2.44)$$

These equations can be solved iteratively by fixing the variable that is not to be minimized. (2.44) can be solved by an exhaustive search in the discrete domain for each pixel. To solve (2.43), it has to be rewritten in a matrix notation as

$$\min_{\mathbf{d}} \{\|\mathbf{AGd}\|_{\varepsilon} + \frac{1}{2\theta}\|\mathbf{d} - \mathbf{a}\|_2^2\} , \qquad (2.45)$$

with $\mathbf{d} \in \mathbb{R}^{MN \times 1}$ and $\mathbf{a} \in \mathbb{R}^{MN \times 1}$ being column vectors of ξ and α and $G \in \mathbb{R}^{MN \times MN}$ being a diagonal matrix containing the pixel weights. M denotes the number of rows of the image and N denotes the number of columns of the image. The matrix $A \in \mathbb{R}^{2MN \times MN}$ is another way of writing the discrete gradient and is defined as

$$A = \begin{pmatrix} B & 0 & \cdots \\ & \ddots & \\ \cdots & 0 & B \\ & C & \end{pmatrix} . \qquad (2.46)$$

$B \in \mathbb{R}^{M \times N}$ is a matrix having 1 at the diagonal and -1 at the entry right of the diagonal. $C \in \mathbb{R}^{MN \times MN}$ is a matrix having 1 at the diagonal and -1 at the elements starting from $(N+1, 1)$ and going diagonal downwards from there on. An example for $M = N = 2$ would be

$$A = \begin{pmatrix} 1 & -1 & 0 & 0 \\ 0 & 1 & 0 & 0 \\ 0 & 0 & 1 & -1 \\ 0 & 0 & 0 & 1 \\ 1 & 0 & -1 & 0 \\ 0 & 1 & 0 & -1 \\ 0 & 0 & 1 & 0 \\ 0 & 0 & 0 & 1 \end{pmatrix} . \qquad (2.47)$$

A^{T} can also be expressed in terms of matrix operators as it is the negative divergence.

To replace the Huber norm, the Legendre-Fenchel transformation is used. This transformation is described in Section 2.1. The Legendre-Fenchel transformation of itself is the original function again, which is also called biconjugate. The biconjugate is defined as

$$f^{**}(\mathbf{x}) = f(\mathbf{x}) = \sup_{\mathbf{p} \in \mathbb{R}^n} \{\mathbf{p}^{\mathsf{T}}\mathbf{x} - f^*(\mathbf{p})\} . \qquad (2.48)$$

The biconjugate of the Huber norm is then

$$f^{**}(\mathbf{x}) = \sup_{\|\mathbf{p}\|_2 \leq 1} \{\mathbf{p}^{\mathsf{T}}\mathbf{x} - \frac{\varepsilon}{2}\|\mathbf{p}\|_2^2\} . \qquad (2.49)$$

The Huber norm in (2.45) can now be replaced by its biconjugate. Then, the equation becomes

$$\min_{\mathbf{d}}\{ \sup_{\|\mathbf{p}\|_2 \leq 1} \{\mathbf{p}^T A G \mathbf{d} - \frac{\varepsilon}{2}\|\mathbf{p}\|_2^2\} + \frac{1}{2\theta}\|\mathbf{d} - \mathbf{a}\|_2^2\} . \tag{2.50}$$

For the supremum, the derivate in terms of $\mathbf{p}$ has to be calculated, resulting in

$$\frac{\partial}{\partial \mathbf{p}}\{\mathbf{p}^T A G \mathbf{d} - \frac{\varepsilon}{2}\|\mathbf{p}\|_2^2\} = A G \mathbf{d} - \varepsilon \mathbf{p} . \tag{2.51}$$

As the supremum is needed, a gradient ascent has to be done:

$$\frac{\mathbf{p}^{n+1} - \mathbf{p}^n}{\sigma_p} = A G \mathbf{d}^n - \varepsilon \mathbf{p}^{n+1} . \tag{2.52}$$

The equation can be rearranged to solve for $\mathbf{p}^{n+1}$ and then becomes

$$\mathbf{p}^{n+1} = \frac{\sigma_p A G \mathbf{d}^n + \mathbf{p}^n}{1 + \sigma_p \varepsilon} . \tag{2.53}$$

As $\|\mathbf{p}\|_2$ has to be less or equal than 1, the equation has to be changed to

$$\mathbf{p}^{n+1} = \Pi_{\mathbf{p}} \left(\frac{\sigma_p A G \mathbf{d}^n + \mathbf{p}^n}{1 + \sigma_p \varepsilon} \right) , \tag{2.54}$$

where $\Pi_{\mathbf{p}}(\mathbf{x}) = \mathbf{x}/\max(1, \|\mathbf{x}\|_2)$. This scales $\mathbf{p}$ to a length of 1 if it is longer and preserves the length if it is shorter.

To do the minimization of the whole term in (2.50), the product $\mathbf{p}^T A G \mathbf{d}$ has to be rewritten. This term is the scalar product $\langle A G \mathbf{d}, \mathbf{p}\rangle$. As the scalar product is commutative, this is the same as $\langle \mathbf{p}, A G \mathbf{d}\rangle$, resulting in $\mathbf{d}^T G^T A^T \mathbf{p}$. For the minimization, differentiation in terms of $\mathbf{d}$ has to be done:

$$\frac{\partial}{\partial \mathbf{d}}\{ \sup_{\|\mathbf{p}\|_2 \leq 1} \{\mathbf{d}^T G^T A^T \mathbf{p} - \frac{\varepsilon}{2}\|\mathbf{p}\|_2^2\} + \frac{1}{2\theta}\|\mathbf{d} - \mathbf{a}\|_2^2\} = G^T A^T \mathbf{p} + \frac{1}{\theta}(\mathbf{d} - \mathbf{a}) . \tag{2.55}$$

As the minimum is searched, a gradient descent has to be done:

$$\frac{\mathbf{d}^{n+1} - \mathbf{d}^n}{\sigma_d} = -G^T A^T \mathbf{p}^{n+1} - \frac{1}{\theta}(\mathbf{d}^{n+1} - \mathbf{a}^n) . \tag{2.56}$$

When this equation is rearranged to solve for $\mathbf{d}^{n+1}$, it becomes

$$\mathbf{d}^{n+1} = \frac{\sigma_d(\frac{\mathbf{a}^n}{\theta} - G^T A^T \mathbf{p}^{n+1}) + \mathbf{d}^n}{1 + \frac{\sigma_d}{\theta}} . \tag{2.57}$$

During the update steps in (2.54) and (2.57), the variable $\mathbf{a}^n$ is fixed.

After $\mathbf{d}^{n+1}$ has been calculated, $\mathbf{a}^{n+1}$ has to be determined. Therefore, $\mathbf{d}^{n+1}$ is fixed and an exhaustive search on

$$E_{\text{aux}}(\mathbf{u}, d_{\mathbf{u}}, a_{\mathbf{u}}) = \frac{1}{2\theta}(d_{\mathbf{u}} - a_{\mathbf{u}})^2 + \lambda C(\mathbf{u}, a_{\mathbf{u}}) + \lambda_2(a_{\mathbf{u}} - d_{i,\mathbf{u}})^2 \tag{2.58}$$

is done to find $\mathbf{a}^{n+1}$ minimizing E_{aux}. The accuracy of $\mathbf{a}^{n+1}$ can be increased by doing a single Newton step

$$\hat{a}_{\mathbf{u}}^{n+1} = a_{\mathbf{u}}^{n+1} - \frac{\frac{\partial}{\partial a_{\mathbf{u}}} E_{\text{aux}}(\mathbf{u}, d_{\mathbf{u}}^{n+1}, a_{\mathbf{u}}^{n+1})}{\frac{\partial^2}{\partial a_{\mathbf{u}}^2} E_{\text{aux}}(\mathbf{u}, d_{\mathbf{u}}^{n+1}, a_{\mathbf{u}}^{n+1})}. \tag{2.59}$$

When $\mathbf{p}^{n+1}$, $\mathbf{d}^{n+1}$ and $\mathbf{a}^{n+1}$ have been calculated, θ is updated as

$$\theta^{n+1} = \theta^n(1 - \beta n) \tag{2.60}$$

and the steps are repeated. This is done as long as $\theta^n > \theta_{\text{end}}$. The scheme is initialized with $\mathbf{p}^0 = 0$, $n = 0$ and $\mathbf{d}$ and $\mathbf{a}$ initialized as described in Section 2.4.2.

Chapter 3

Implementation

This chapter describes some important aspects of the implementation of the software. The software is implemented in C++ under the usage of some libraries and tools that are listed in Table 3.1.

First, the configuration is introduced as it is used throughout the whole software. The capter continues with an introduction to a CUDA framework that has been implemented in order to simplify programming of computations on the GPU. Afterwards, the sparse reconstruction is being discussed. In the end, some aspects of the implementation of the dense reconstruction are shown.

3.1 Configuration

All configuration parameters are held by the class *Configuration*. It is implemented as singleton to make sure that there is only one valid configuration throughout the whole software [5]. The implementation as singleton also has the advantage that the instance can be reached easily from everywhere in the software without needing to pass it to each function or constructor as a parameter. The class diagram of the class *Configuration* is given in Figure 3.1.

The configuration parameters are saved in INI files. These files can be loaded via the method *addIniFile*. It is possible to load more than one INI file. In this case, the files are searched in reverse order for a parameter. That means the last added INI file will be searched for the parameter first. Set parameters are saved to the last added INI file, regardless of their origin INI file. If a parameter can not be found in any of the INI files, it is searched in the default configuration. This is an INI file that is compiled into the software holding default values for all parameters.

For some tasks, it is necessary to set parameters without saving them to the INI files. Therefore, a temporary layer can be created above the INI files. This is done with the method *createTemporaryLayer*. All changes are then saved to this temporary layer while parameters that have not been set in the temporary layer are still loaded from the INI files. The temporary layer can be deleted via *deleteTemporaryLayer*, discarding all changes made to it.

The parameters are organized in a two-layer hierarchy. There are groups containing the parameters. This is normal for INI files. In the method signatures, the group is named *group* and the parameter itself is called *entry*. The methods of the class *QSettings* expect group and parameter name in one string, delimited by a special character. The method *getEntryName*

Name	Function	URL
OpenCV	Image processing	http://www.opencv.org/
Qt	GUI	http://www.qt-project.org/
CUDA	Computations on GPU	https://developer.nvidia.com/cuda-zone/
Boost	Utility functions	http://www.boost.org/
SiftGPU	Feature tracking on GPU	http://cs.unc.edu/ ccwu/siftgpu/
pba	Bundle adjustment	http://grail.cs.washington.edu/projects/mcba/
sba	Bundle adjustment	http://users.ics.forth.gr/ lourakis/sba/
CMake	Configure compilation	http://www.cmake.org/
Doxygen	Documentation	http://www.stack.nl/ dimitri/doxygen/

Table 3.1: Libraries and tools that are used in the software. Also the functions and URLs of the projects are shown. The URLs were checked on 9th October 2014.

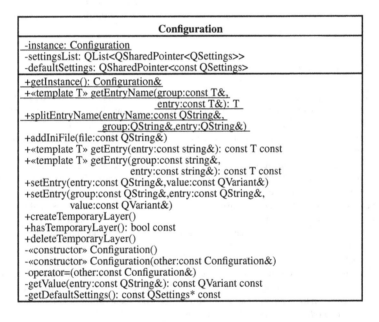

Figure 3.1: Class diagram of the class *Configuration*.

takes the group and parameter name and constructs this string. The other direction is done by the method *splitEntryName*. It takes a string containing group and parameter name and splits it into the parts. The method *getEntry* for getting a parameter value also expects a template parameter. This denotes the type the parameter value should be casted to.

3.2 CUDA Framework

CUDA is a technique developed by NVIDIA in order to execute parallel computations on graphics processing units (GPUs) manufactured by NVIDIA. There are bindings for several languages. This also includes C, which is used here. The computations are done in functions called kernels that are scheduled in parallel on the GPU. The kernels are arranged in blocks. Each kernel does the computation for one single pixel. The kernel is able to ask for the ID of the block as well as the ID of the kernel and also for block sizes, so it is able to determine which pixel should be calculated.

During compilation, CUDA files have to be compiled by the CUDA compiler. In order to separate them from normal C/C++ files, they get the extension *.cu*. The CUDA compiler generates object files which are then normally linked into the program during the linking phase. CUDA extends the C language by some API functions and some syntax in order to control the parallel execution. These extensions are not available in files that are compiled by the normal C/C++ compiler. Therefore, the preprocessor macro __CUDACC__ can be checked. It is only defined during compilation by the CUDA compiler.

The kernels are executed on the GPU. As the GPU has separate memory and is not able to access the main memory[1], doing computations on the GPU seems to be complicated. First, memory has to be allocated on the host (main memory) and on the device (memory on the graphics card). Then, the host memory has to be filled with the input data. Afterwards, the content of the host memory has to be transferred to the device memory. The device then does all computations on the device memory and the data has to be copied back to the host memory subsequently. In the end, the allocated memory has to be freed on host and device. Because of this laborious process, it is important to check carefully whether a computation on the GPU gives performance benefits. For small calculations, it can be faster to execute them on the CPU.

The allocated memory has to be freed manually by the programmer. This has to be done before the function is exited and the pointer to the memory is destroyed. It can be complicated to find all exit points of a function, especially in case of exceptions. Of course, it can also be simply forgotten. Inspired by the book 'Effective C++' from Scott Meyers [16], some smart pointer like classes have been developed to solve this problem. There are three different classes for three different types of memory, namely *HostPinnedMemory*, *DeviceMemory* and *DeviceMemory2D*. Their class diagrams can be found in Appendix A. The classes are implemented as templates in order to support different data types.

Blocks in the main memory are represented by *HostPinnedMemory* (Figure A.3). There is also the addition that these blocks have to stay in the main memory and can not be swapped by the operating system. Such memory can be used for copy operations between main memory and graphics card that are executed asynchronously to CPU operations. The memory can be

---

1 In fact, the GPU is able to access the main memory since version 2.2 of the toolkit but this has some limitations and performance disadvantages.

allocated via the constructor that takes the number of elements as a parameter or via the method *allocate*. The memory is freed by the method *free* or the destructor. Instances of the class can be converted to pointers of the template parameter in order to be used in C interfaces.

Memory on the graphics card is represented by two different classes, depending on the type of usage. The class *DeviceMemory* (Figure A.4) represents linear memory and the class *DeviceMemory2D* (Figure A.5) represents two-dimensional memory. Threads in CUDA are composed in so called warps. Currently, there are 32 threads inside a warp. All threads of a warp share the same program counter and therefore the same execution path, but there is a limitation on how to access memory. Read operations from the memory have also to be done simultaneously by all threads in a warp and it is only possible to access blocks at an address that is a multiple of the warp size in one reading operation. If the address does not start at a multiple of the warp size, more than one reading operation has to be performed. Therefore, the class *DeviceMemory2D* arranges the memory in such a way that new lines always start at a multiple of the warp size. The class *DeviceMemory* has an interface similar to *HostPinnedMemory* besides the fact that it supports copying. The support for copying is necessary as the instances have to be copied to the graphics card when a kernel is called. When the instance is copied, a flag in the new instance is set indicating that this is only a copy. The method *free* and the destructor do nothing when called on an instance that is only a copy. This procedure ensures that the memory is only freed when the original instance is deleted. The class *DeviceMemory2D* implements the same method for copies. It does not support to be directly converted to a pointer of the template type but therefore has the method *data*. It also implements the array operator that gives a pointer of the template type to a given row.

It is not only possible to parallelize the computation on one image. It is also possible to do different computations and memory transactions in parallel, but this will not be done by default. In order to enable this parallelism, the memory transactions and computations have to be done in different streams. Each stream will be executed sequentially, but different streams can be executed in parallel. Construction and destruction of the CUDA streams are encapsulated in the class *CudaStream* (Figure A.2). The stream is created in the constructor and deleted in the destructor. Instances of the class can be converted to the C structure expected by the CUDA functions. The method *join* blocks the current thread until all operations in the stream have been finished.

The last class in the framework is *CudaException* (Figure A.1). It is a child class of the class *exception* from the standard library. The constructor with the parameter of the type *cudaError_t&* is only available during compilation by the CUDA compiler. It retrieves the text belonging to the error code from the CUDA API and stores it inside the member variable.

Besides the classes, there are also some functions in the CUDA framework. One function to be mentioned is the function *copy*. This function has several overloads to copy data between all memory classes of the framework and to copy data between *HostPinnedMemory* and the OpenCV class *Mat*.

3.3 Sparse Reconstruction

The sparse reconstruction with the Structure from Motion approach consists of several steps. These are shown in Figure 2.1. For the implementation of the sparse reconstruction, the strategy pattern is used [5]. Each step of the reconstruction is a strategy and can therefore be easily exchanged by another method. The configuration controls which algorithm is used for each strategy. Figure B.1 shows an overview of the classes used for the implementation. Only the base classes of the strategies are shown, not the concrete implementations.

3.3.1 Structure from Motion

The sparse reconstruction is controlled by the class *SfM*. Figure B.2 shows the class diagram. An instance of the class has to be constructed with the matrix K, representing the intrinsic camera parameters. For each frame, the method *nextFrame* has to be called. The instance of the class *SfM* then does the whole reconstruction process for this frame and returns whether this frame was considered as being a keyframe. After the last frame in the sequence, the method *finish* has to be called. It ensures that all frames up to the last frame have been taken into account for the reconstruction. The positions of the cameras can be retrieved via the method *getProjectionMatrices* and the method *getPointCloud* returns all reconstructed 3D points. With the method *exportData* it is possible to export the current data of the *SfM* instance for debugging purposes.

The class *SfM* uses the class *Trajectory* for maintaining the trajectories found during tracking. A trajectory consists of an ID, a 3D point that can be valid or invalid and the index of the 2D point for each frame that includes the 3D point. The ID is set automatically at the creation of an instance and can not be changed. It is used to find the right trajectory again after some calculations, for example after bundle adjustment. After creation of the instance, the 3D point is invalid. It becomes valid when the 3D point is set. It can be invalidated by the method *invalidatePoint* in case the 3D point is filtered out by a 3D filter. New frames are added to the trajectory via the method *addFramePoint*. After filtering, frames can be removed with the method *removeFrame*. The point for a given frame can be retrieved via the method *getFramePoint*. This method will return the value -1 if the 3D point is not visible in the frame. Trajectories are deleted if the number of 2D points is too low or if they include only one point and this frame is too far away in the sequence from the current frame. Therefore, the number of 2D points and the last frame can be retrieved with the methods *getNumberOfPoints* and *getLastFrame*.

3.3.2 Tracking

The tracking strategy is responsible for finding keypoints in the images, extracting the descriptors and matching the descriptors of different frames in order to find correspondences. Figure B.3 shows the class diagram of the tracking strategy.

Matching for a new frame is done by the method *computeNextFrame*. It gets the new frame as input parameter and outputs the keypoints that were found as well as the matches of the keypoints with the keypoints in previous frames. The previous frames, their keypoints and descriptors are stored inside the instance of *Tracking*. The matches have already been filtered during the method. Figure B.1 shows that filtering and matching are also implemented as strategies. Only one strategy for each of them has been implemented. Brute force matching has been implemented as a matching strategy. In this strategy, all descriptors of one frame are checked against all descriptors of the other frame to find the one with the lowest distance. It is implemented for CPU and GPU. As there can be wrong matches, filtering has to be performed after matching. Schlüter et al. have examined different types of filters and shown that a K Nearest Neighbours Check performs best in terms of precision and recall [22]. An additional Angular Consistency Check increases the precision but also decreases the recall. As both of these effects are marginal, only the K Nearset Neighbours Check has been implemented. It checks if more than half of the K neirest neighbours of a keypoint in one frame are equal to the K neirest neighbours of the corresponding keypoint in the other frame.

The tracking can be executed on the CPU or the GPU. As different OpenCV data types have to be used depending on this choice, an additional layer consisting of the classes *TrackingCpu* and *TrackingGpu* has been introduced. These classes encapsulate the steps that have to be done before and after matching. The algorithms SIFT and SURF can be used for finding keypoints and extracting the descriptors. For executing SIFT on the GPU, the library SiftGPU is used [27]. As this library has a different interface than the OpenCV classes and also has its own matcher, the class controlling the library is not inheriting *TrackingCpu* or *TrackingGpu*.

The class method *showMatches* can be used for debugging purposes. It gets the two images as well as the keypoints in each image. These have to be in the same order in both vectors, corresponding to each other. Then, both images are displayed beside each other and the keypoints will be marked and connected. The resulting image is shown in a tab in the main window with the title given as the last parameter.

3.3.3   Pose Estimation

The pose estimation strategy is responsible for finding the positions of the cameras. It estimates the essential matrix and the resulting extrinsic camera parameters between two frames up to scaling. The class diagram is shown in Figure B.4.

The class *PoseEstimation* is implemented as a functor. The function-call operator takes the corresponding points from both frames and returns the essential matrix as well as the extrinsic camera parameters. The position of the first camera is always considered as being in the origin and the position of the second camera is given relative to the first camera. The function-call operator also returns the indicies of the points that are considered to be outliers. The essential matrix is only estimated up to scale. Therefore, scale estimation has to be done to integrate the newly estimated camera position correctly into the scene. Scale estimation can be done in two ways. One way is to do bundle adjustment and keep all previous frames as well as the 3D points fixed and only change the new camera position. The other way is to do an explicit scale estimation. This can be done with the method *robustScaleEstimation*. Non-quantitative tests have shown that bundle adjustment performs much better in terms of robustness. Therefore, only bundle adjustment has been used for the evaluations.

Classes for estimating the essential matrix by using the five-point and eight-point algorithm have been implemented. Both use a RANSAC scheme to do a robust estimation. The class *Ransac5Point* uses the five-point algorithm and the class *Ransac8Point* uses the eight-point algorithm.

3.3.4   Triangulation

The triangulation strategy is responsible for backprojecting the corresponding 2D points of two frames into the 3D space. The strategy is implemented as a functor and the class diagram shown in Figure B.5.

The function-call operator takes the essential matrix as well as the extrinsic camera parameters for both cameras as input parameters. It also gets the corresponding 2D points. It returns the backprojected 3D points in the output parameter *result*. The only implementation of the strategy, the class *TriangulationOpenCV*, uses the OpenCV function *triangulatePoints* to execute the triangulation. Before the triangulation itself can be done, the positions of the 2D points have to be corrected like described in Section 2.3.3. This is done by the OpenCV function *correctMatches*. After the correction, the points are backprojected.

3.3.5   Bundle Adjustment

The last step in the sparse reconstruction is to optimize the camera positions and the positions of the 3D points. This is done by the bundle adjustment strategy. Figure B.6 shows the class diagram of this strategy.

The constructor of the class *BundleAdjustment* takes the intrinsic camera parameters as well as the instance of the class *SfM* to operate on as input parameters. The hierarchy of the bundle adjustment strategy is declared as friend of the *SfM* class in order to manipulate the trajectories which are private in the class *SfM*. The bundle adjustment task is executed via the method *doBundleAdjustment*. It gets the index of the first frame and the last frame to be used for bundle adjustment. It is possible to fix a number of frames with the parameter *fixedFrames*. The fixed frames are always taken from the beginning of the sequence of frames to operate on. With the parameter *fix3DPoints* it is possible to fix the 3D points. The last two parameters are used when bundle adjustment is taken for scale estimation.

The strategy does not only execute the bundle adjustment, but also does some filtering. Bundle adjustment is not robust to outliers. Therefore, before bundle adjustment is executed, the 3D points are filtered. The median of the distance to the K nearest neighbours is estimated for all 3D points. Then, a threshold for the maximum distance is estimated by taking the median of these medians and multiplying it with a factor. All 3D points that have a higher median distance to the K neirest neighbours than the threshold are considered to be outliers and are deleted. After bundle adjustment has been executed, a filtering on the 2D points is done. The 3D points are projected into all frames where they occur and the resulting point is compared to the 2D point of the trajectory belonging to the frame. If the geometric error is higher than a threshold, the 2D points are deleted from the trajectories. In the end, trajectories with too few 2D points are also deleted. The process of bundle adjustment and 2D filtering is repeated until no 2D points have been deleted or a given number of iterations is reached.

There are two implementations of the bundle adjustment strategy that use different libraries for bundle adjustment. The first one is the class *Sba* that uses the *sba* library [11]. The second one is the class *Pba* that uses the *pba* library [28]. The library *pba* executes the bundle adjustment on the GPU and is therefore much faster than *sba*, but *sba* is much more robust than *pba* and *pba* often diverged during non-quantitative tests.

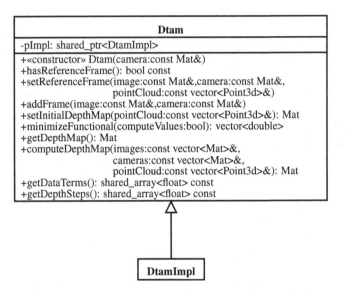

Figure 3.2: Class diagram of the dense reconstruction. Only the public interface of the class *Dtam* is shown.

3.4 Dense Reconstruction

The dense reconstruction has to be done with some knowledge of the sparse reconstruction. Possible scenarios are to execute the sparse reconstruction for the whole scene and then execute the dense reconstruction or to execute the dense reconstruction parallel to the sparse reconstruction. Both methods can be done with the class *Dtam*. Most parts of the dense reconstruction can be executed in parallel for each pixel. Therefore, these parts are executed on the GPU. To be able to do the dense reconstruction in parts of the software which are not compiled by the CUDA compiler, the interface is decoupled from the implementation by the bridge design pattern [5]. The interface is declared in the class *Dtam* and the implementation is defined in the class *DtamImpl*. Figure 3.2 shows the class diagram of the dense reconstruction.

The first step in the dense reconstruction is to set a reference frame. This allocates the necessary memory, converts the reference frame into the $r\phi\theta$ color space, calculates the pixel weights and determines the samples in the inverse depth space for calculating the cost volume. Therefore, the current point cloud is needed and given to the method *setReferenceFrame* as the parameter *pointCloud*. The problem is that the depth range of the whole scene is not known at the beginning of the reconstruction. Only the depth range of the first frames is known. If the range is taken too small, some values are cut off. If the range is taken too large, the samples are far away from each other resulting in a bad resolution. In addition, some samples will not be used. In order to weaken this problem, the inverse depth range of the first frames is taken and some fraction of it is added before and after the range.

```
1  __global__ void sphericalDataTerm(DeviceMemory2D<unsigned char> image,
2                                      DeviceMemory2D<float> dataTerms,
3                                      int width, int height) {
4      // get x- and y-coordinate to handle
5      int x = threadIdx.x + blockIdx.x * blockDim.x;
6      int y = threadIdx.y + blockIdx.y * blockDim.y;
7      // only calculate parts of the image
8      if ((x < image.getWidth()) && (y < image.getHeight())) {
9          // get pointer to b, g and r value of pixel
10         unsigned char *pixel = &(image[y][3 * x]);
11         // get r, g and b value
12         float b = pixel[0];
13         float g = pixel[1];
14         float r = pixel[2];
15         // get pointer to data term
16         float *dataTerm = &(dataTerms[y][2 * x]);
17         // calculate theta
18         if (r == 0)
19             dataTerm[0] = PI_2;
20         else
21             dataTerm[0] = atan(g / r);
22         // calculate phi
23         if (b == 0)
24             dataTerm[1] = PI_2;
25         else
26             dataTerm[1] = asin(sqrt(r * r + g * g) /
27                               sqrt(r * r + g * g + b * b));
28     }
29 }
```

Listing 3.1: Conversion from RGB to $r\phi\theta$ color space.

Listing 3.1 shows the source code to convert an image from the RGB color space to the $r\phi\theta$ color space. It will be executed on the GPU and also shows some aspects of CUDA programming. The first thing to mention is the keyword __global__ in line 1. It declares the function to be executed on the GPU and to be callable from CPU and GPU code. The function gets the input image in the parameter *image* and the width and the height of the image in the parameters *width* and *height*. The result will be stored in the parameter *dataTerms*. Each thread is responsible for calculating one pixel of the image. Therefore, each thread receives the information which pixel has to be calculated. This is shown in lines 5 and 6. As mentioned in Section 3.2, threads are organized in a two-dimensional array called block. The blocks are also organized in a two-dimensional array called grid. It is possible to get the index of the current thread in x- and y-direction as well as the index of the current block and the block dimensions. With these values it is possible to determine exactly which pixel has to be calculated, independently of the size of the blocks. All blocks have the same dimensions, so there can be threads which calculate parts that are not inside the image borders. Line 8 checks if the current pixel is inside the image borders. The rest of the listing is standard C code that calculates the transformation like in (2.18). OpenCV stores the RGB components in a different order, namely BGR. This is respected in the lines 12 to 14.

After the reference frame has been set, new frames can be added in order to iteratively compute the cost volume. This can only be done after a new keyframe in the sparse reconstruction has occured because camera positions are only estimated for keyframes. Therefore, after each keyframe all frames after the last keyframe can be added to the dense reconstruction via the method *addFrame*. Each frame is converted into the $r\phi\theta$ color space and then the cost volume is updated. Besides the cost volume, there is also a reprojection counter for all voxels of the cost volume. With this reprojection counter it is possible to have a normalized cost volume after each added frame.

An inverse depth map has to be created from the point cloud in order to couple the dense reconstruction to the results of the sparse reconstruction. This is done by the method *setInitialDepthMap*. All 3D points of the point cloud are projected into the image that represents the inverse depth map and the resulting pixels are set to the inverse depth of the 3D point. After all 3D points have been projected into the image, all other pixel values are set to the value of the nearest neighbour in the inverse depth map.

The last step in the dense reconstruction is to minimize the functional from (2.40) like described in Section 2.4.4. This minimization can be done in parallel for each pixel of the inverse depth map. It is therefore executed on the GPU. The matrix A is replaced by the discrete gradient and the matrix $-A^T$ is replaced by the discrete divergence. Both are defined by Wedel and Cremers in [25]. The discrete gradient $(\nabla u)_{i,j} = ((\nabla u)^1_{i,j}, (\nabla u)^2_{i,j})^T$ at coordinate (i,j) is defined for images with width N and height M as

$$
(\nabla u)^1_{i,j} = \begin{cases} u_{i+1,j} - u_{i,j} & \text{if } i < N, \\ 0 & \text{if } i = N, \end{cases} \text{ and}
$$

$$
(\nabla u)^2_{i,j} = \begin{cases} u_{i,j+1} - u_{i,j} & \text{if } j < M, \\ 0 & \text{if } j = M. \end{cases} \tag{3.1}
$$

The discrete divergence is defined as

$$
(\text{div}\mathbf{p})_{i,j} = \left(\text{div} \begin{bmatrix} p^1 \\ p^2 \end{bmatrix} \right)
$$

$$
= \begin{cases} p^1_{i,j} - p^1_{i-1,j} & \text{if } 1 < i < N \\ p^1_{i,j} & \text{if } i = 1 \\ -p^1_{i-1,j} & \text{if } i = N \end{cases} + \begin{cases} p^2_{i,j} - p^2_{i,j-1} & \text{if } 1 < j < M, \\ p^2_{i,j} & \text{if } j = 1, \\ -p^2_{i,j-1} & \text{if } j = M. \end{cases} \tag{3.2}
$$

All steps of the dense reconstruction can be executed at once with the method *computeDepthMap*. This method calls all previously described methods in the right order to compute an inverse depth map.

Chapter 4

Evaluation

This chapter gives an evaluation of the implemented 3D reconstruction algorithm. It starts with introducing the ground truth data that has been used for all evaluations. Afterwards, the sparse reconstruction and then the dense reconstruction are evaluated. In the end, selected performance test are shown.

4.1 Ground Truth Data

For the evaluation of methods, ground truth data is needed. The proposed approach should be used to reconstruct liver surfaces from mini-laparoscopic sequences. A quantitative evaluation on clinical data would therefore be the best. In order to get ground truth data for sequences of real livers, some additional examinations like a CT scan have to be done. Unfortunately, this is not done at the moment and it is therefore not possible to do quantitative evaluations on clinical data. Another possibility to evaluate the method has to be found.

15 different sequences with their corresponding ground truth data have been recorded in the TUVision institute in order to evaluate reconstruction methods. The sequences consist of five different types of camera motions that are listed in Table 4.1. Each of them is executed on three different phantoms that have different types of cirrhotic nodules.

The camera motion was tracked during the recording of the sequences with a tracking system. The phantoms have been measured with the same tracking system. Therefore, the surface has been scanned with a tool that has been tracked by the tracking system. Afterwards, a mesh has been created from the tracking data. The position of the phantom in the coordinate system of the tracking system has also been estimated.

The procedure of getting the ground truth data has also some problems. First, image capturing of the camera and position measuring of the tracking system are not synchronized. Therefore, the tracked positions are not the exact positions of the camera at the time of image capturing. Petersen et. al worked on the synchronization of image capturing and position measuring [20]. Another problem is the point where the camera position is measured. A tool is mounted at the camera and tracked by the tracking system. The tip of the endoscope containing the image plane is 50 cm apart from the tool. Small errors in the measurements of the camera positions therefore result in much bigger errors for the evaluation as the method works on the images. Mejía et al. worked on the hand-eye calibration [15].

Sequence	Description
Sequence 1	Translational sidewards motion along the phantom
Sequence 2	Translational forward motion towards the phantom
Sequence 3	Small circular motion close to the phantom
Sequence 4	Big circular motion far away from the phantom
Sequence 5	Small circular motion far away from the phantom

Table 4.1: Overview of camera motions in ground truth sequences. These motions were performed during recording of the sequences.

4.2 Sparse Reconstruction

It is necessary to evaluate the sparse reconstruction for itself as its results are needed by the dense reconstruction. For the evaluation, the sparse reconstruction is executed on all 15 sequences. The results are compared to the ground truth data. The estimated camera positions are registered to the ground truth camera positions. Afterwards, the translational and rotational errors in the camera positions are estimated. For each frame, the positions of the visible 3D points are evaluated. Therefore, first the corresponding 2D points are taken and backprojected into the 3D space. Ground truth depth maps provide the depth values for the 3D points. Then the estimated 3D point cloud is registered to the ground truth point cloud and a scale factor is estimated and the point cloud is scaled. The resulting point cloud is then compared to the ground truth point cloud. Appendix C contains the results for the evaluation of the sparse reconstruction. The values on the abscissa describe the frame index beginning with 1 at the first frame that has ground truth data available.

The strategies that are used during the sparse reconstruction are *SiftGpu* for tracking, *Ransac-5Point* for pose estimation with bundle adjustment for scale estimation, *TriangulationOpenCV* for the triangulation and *Sba* for bundle adjustment. A new keyframe is assumed if the median distance of the feature points between two frames is 50 pixels or higher and if there are 20 or more correspondences.

Sequence 3 of Phantom 2 and Sequences 1, 2 and 3 of Phantom 3 have problems during the reconstruction. In Sequence 1 of Phantom 3, a problem with scale estimation occurs. The 26th frame is considered as being a keyframe and bundle adjustment is performed after pose estimation and triangulation. During this bundle adjustment step and its filter routines, all 3D points are discarded. The problem has to be examined in the future. The other three sequences have problems with the keyframe estimation. The conditions of the minimum distance and minimum number of the correspondences both have to be fulfilled for a new keyframe. As these thresholds are fixed, it can be possible that no new keyframe is estimated. The last frame in a sequence has to be a keyframe because pose estimation and triangulation is only done between keyframes. If there are no reconstructed 3D points that are visible in the last frame, the camera positions for all cameras from the last keyframe up to the last frame of the video have to be extrapolated. Therefore, it is assumed that the motion from the frame preceding the last keyframe to the last keyframe is the same as between all frames from the last keyframe to the last frame of the sequence. Camera positions are created using this motions for all missing frames. Afterwards, bundle adjustment is performed that only varies the positions of these cameras.

In the end, a normal bundle adjustment step is executed varying all camera positions and 3D points. The evaluation results of these three sequences show that this procedure is not optimal. Especially in the rotation error, it is obvious to see at which frame the tracking was lost. No 3D points have been triangulated for the frames after the tracking has been lost. The extrapolated camera positions have been so bad that the points in the trajectories for these cameras have also been deleted during 2D filtering. Therefore, no 3D points are visible in the extrapolated frames and the point cloud error could not be estimated for these frames. The point cloud error for the frames before the tracking has been lost is in the same dimensions as the point cloud error for sequences where the reconstruction had no problems. The evaluation of Sequence 3 of Phantom 2 lost the tracking at around 20 frames. For the results in Figure C.8 the minimum number of correspondences was reduced to 15. This also shows that the keyframe estimation is an important topic to work on.

All other sequences show good results. Except Sequence 4 of Phantom 1, the median point cloud error is below 1 mm for nearly all frames. The Structure from Motion approach works iteratively over all frames. This can be seen in some of the error plots of the camera positions. There, the error is increasing over the sequence. On the first look, it might seem strange that the point cloud error is less than the error of the camera positions. This effect is caused by the method of the evaluation. As the ground truth 3D points are created by using a ground truth depth map for the frame and then the estimated points are registered to them, the orientation of the camera is not important for the estimation of the point cloud error. The errors in the camera positions also do not necessarily have an impact on the dense reconstruction. The amount of the influence depends on the direction of the position error and on the image content. If the camera is shifted along the optical axis, there will be no problem. There will also be no problem if there are vertical lines in the image and the camera is shifted vertically. These are only two examples when there will be no influence by the error in the camera position but it is important to keep the evaluation results of the sparse reconstruction in mind when evaluating the dense reconstruction. It is even more important to keep in mind the amount of sparse reconstructions that did not work and are not able to give a reliable input to the dense reconstruction.

4.3 Dense Reconstruction

The evaluation of the dense reconstruction is executed with the same parameters for the sparse reconstruction as in Section 4.2. The sparse reconstruction for Sequence 3 of Phantom 2 is also executed with a minimum of 20 correspondences for the keyframe estimation. The dense reconstruction is executed on each 10th frame and the directly following 29 frames in the sequence are used for the computation of the cost volume. The inverse depth map is initialized with the minimum of the data terms and the interpolated inverse depth map from the sparse reconstruction is coupled to the solution via λ_2. The sparse reconstruction is also executed only on these 30 frames. After the inverse depth map has been computed, a scale factor between this inverse depth map and the ground truth inverse depth map is estimated. A least squares method is used in the depth space to estimate this scale factor. The scaled depth map is then compared to the ground truth depth map. For the estimation of the scale factor and the comparison of the depth maps, the depth space and not the inverse depth space is used. Therefore, the errors will be given in millimeters.

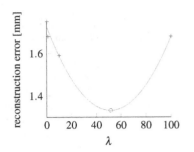 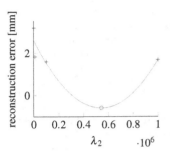

Figure 4.1: Estimation of dense reconstruction parameters. The dense reconstruction is executed with different parameter values. The crosses indicate these measurements. The line shows a parabola that is estimated to fit the data. The circle marks the minimum of this parabola that is also taken as the parameter value.

Before the dense reconstruction can be evaluated, optimal values for the parameters λ, λ_2 and θ have to be estimated. Therefore, the reconstruction is executed with different values for these parameters on all sequences. It is important that the minimum error over all reconstructions lies inside this value space. Afterwards, a parabola is estimated to fit the data. The minimum of this parabola indicates the optimal value for the parameter. Figure 4.1 illustrates the estimation of the parameter values.

After the optimal parameter values have been estimated, the dense reconstruction can be executed on all sequences with these parameter values to get the final evaluation results. Table 4.2 shows the results of the final evaluation. The median error over all sequences is 2.96 mm. For the individual sequences, the median error ranges from 1.38 mm to 8.6 mm. The reconstructions of Phantom 3 are worse than the reconstructions of the other phantoms. Phantom 3 has a finer surface with smaller nodules and it also has a greater depth range than the other phantoms. The parts of the phantom that are far away from the camera are most of the time very dark. Therefore, it is hard to use these parts for the reconstruction. When looking at Phantoms 1 and 2, one can see that Sequences 3 and 4 perform best. These are a small circular motion close to the phantom and a big circular motion far away from the phantom. Circular motions are better for the dense reconstruction as they give more information about the content in the reference frame. They show nearly the same scene of the reference frame in the following frames but from different viewports. In Sequence 1, the translational sidewards motion, the content of the reference frame is only visible for some frames and afterwards the new frames do not give any information about the reference frame. The frames following the reference frame in Sequence 2, the translational motion towards the phantom, also do not give new information about the reference frame. They have the same viewport as the reference frame. Sequence 5 is slightly worse than the other circular motions. In this sequence, a small circular motion is performed far away from the phantom. This leads to small baselines in the images. The scene is also much darker as the light source is mounted at the tip of the mini-laparoscope and therefore far away from the phantom.

Phantom	Sequence	Percentiles [mm]					
		p_{25}	p_{50}	p_{75}	p_{90}	p_{95}	p_{99}
	1	1.82	2.87	5.03	13.81	20.25	29.85
	2	1.45	3.71	10.80	20.73	25.32	34.47
1	3	0.60	1.38	3.04	5.14	8.30	19.56
	4	0.77	1.74	3.76	8.01	12.48	24.26
	5	1.73	2.62	3.73	8.79	17.00	25.20
	1	1.24	2.86	4.72	11.25	15.42	30.79
	2	2.60	3.71	7.13	12.05	17.23	36.49
2	3	0.73	1.73	4.63	9.16	12.29	20.82
	4	0.75	1.66	3.64	8.54	13.50	28.44
	5	1.23	2.28	4.49	8.96	13.49	27.96
	1	1.04	2.40	5.52	8.49	10.80	24.13
	2	3.63	8.60	13.39	21.18	27.42	54.16
3	3	3.83	6.94	9.75	15.52	28.46	74.57
	4	3.36	5.28	7.75	13.32	24.73	54.45
	5	5.58	7.02	8.83	18.22	34.08	62.11
all	all	1.21	2.96	6.69	11.79	17.70	38.12

Table 4.2: Evaluation results of the dense reconstruction. For each sequence, the percentiles of the reconstruction error are given in millimeters. Also the percentiles of the reconstruction error for all sequences of all phantoms are given.

Figure 4.2 shows an example for the dense reconstruction. Figure 4.2 (a) shows the reference frame. The ground truth for the inverse depth map is shown in Figure 4.2 (b) and the reconstruction is shown in Figure 4.2 (c). The error for each pixel in millimeters is illustrated in Figure 4.2 (d). The interpolated inverse depth map created from the sparse reconstruction has many combs in it because of the nearest neighbour interpolation. These combs are still present in the reconstructed inverse depth map. The coupling to the interpolated inverse depth map therefore seems to be too strong. The problem is that the reconstruction becomes very noisy in case of bad data terms and then needs a strong coupling to the interpolated inverse depth map. A decreasing coupling and the method of initializing the inverse depth map with the interpolated one and then not coupling it later in the process ($\lambda_2 = 0$) have to be examined as they seem to be very promising to solve the problems of the initialization and the strong coupling.

The highest errors appear at edges and in dark regions. It is difficult to reconstruct the scene at edges as there is normally not much information about the region behind these edges. The problems in dark regions appear because of using the $r\phi\theta$ color space. In this color space, colors are represented by the radius r and the two angles ϕ and θ. A small radius corresponds to a dark color and the $r = 0$ denotes black. In dark regions, small changes of the color can lead to high changes in the angles ϕ and θ. This leads to the high errors in dark regions.

Cirrhotic nodules are classified as micronodular (< 3 mm), nodular (3-7 mm) and macronodular (> 7 mm). With a median error of 2.96 mm it is possible to see most of the macronodular cirrhotic nodules in the reconstruction. Even though the overall median error could lead to the opinion to also be able to see nodular cirrhotic nodules in the reconstruction, there are also re-

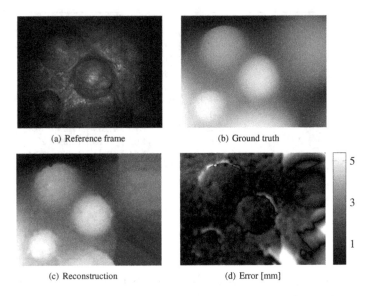

(a) Reference frame   (b) Ground truth

(c) Reconstruction   (d) Error [mm]

Figure 4.2: Example of a reconstructed frame. (a) shows the reference frame. (b) shows the ground truth inverse depth map and (c) shows the reconstructed inverse depth map. The error for each pixel in millimeters is illustrated in (d).

constructions with much higher errors. In some of them it would even be hard to correctly see all macronodular cirrhotic nodules.

4.4 Performance

As one long term goal is to be able to do the reconstruction in real-time, a performance evaluation has to be done. For this evaluation, 30 frames beginning with the frame with index 30 of Sequence 3 of Phantom 1 have been used. The reconstruction has been executed three times with the same parameters as in Section 4.3. The system to do the performance evaluation contains an Intel Core i5-4670 and a NVIDIA GeForce GTX 750 Ti. Table 4.3 gives the results of the performance evaluation. It shows the total reconstruction times in seconds as well as the fraction of the different steps in the reconstruction. The results for all three executions are presented. Steps that take very few time, like the keyframe estimation, are omitted in the table.

   When looking at the total reconstruction time, one can see that the current implementation is far away from real-time. For a sequence of 30 images, a reconstruction could only be called real-time or nearly real-time if it takes 2 seconds or less. Reconstruction times of around 55 seconds are therefore much too long. Looking at the fractions of the sparse and the dense reconstruction, one can see that the dense reconstruction takes only about 16% of the reconstruction time. This

		1	2	3
Sparse Reconstruction	Tracking [%]	14.40	14.44	14.21
	Pose Estimation [%]	5.97	6.02	5.82
	Triangulation [%]	0.55	0.59	0.53
	Bundle Adjustment [%]	55.81	55.32	56.43
Dense Reconstruction	Cost Volume Update [%]	0.57	0.57	0.56
	Initial Depth Map Creation [%]	11.33	11.96	11.52
	Minimization [%]	4.33	4.36	4.28
	Total [s]	55.55	54.88	56.75

Table 4.3: Results of the performance evaluation. For the evaluation, 30 frames starting from frame with index 30 of Sequence 3 of Phantom 1 were used. The measurements were executed three times.

means that the dense reconstruction takes about 9 seconds. This is of course still too much to be considered real-time but the implementation of the dense reconstruction is not optimized so far. The initial depth map creation takes most of the time of the dense reconstruction. This step is executed on the CPU at the moment. Implementing this step on the GPU would lead to an increasing performance. The other steps of the dense reconstruction can also be optimized. There are different types of memory on the graphics card. At the moment, only the global memory is used as it is much easier for the implementation. This memory has a high latency. There is also memory local to a block of threads with very low latency. Therefore, it is better to copy the data that is needed by all threads of a block first to the local memory and then do the calculations.

The fractions of the sparse reconstruction steps show that tracking and bundle adjustment take most of the reconstruction time. Bundle adjustment even takes more than half of the whole reconstruction time. This fraction would increase if the reconstruction will be executed with more frames. The other step that takes very long is tracking. This includes finding SIFT keypoints and features, matching them and maintaining the trajectories. The fact that the sparse reconstruction takes more than 75% of the total reconstruction time shows that this is the part that is not being able to be used in real-time. In the future, the tracking shall be executed in the dense domain replacing the sparse reconstruction but even then the sparse reconstruction has to be used for the initialization of the dense reconstruction and to recover the dense reconstruction when the tracking has been lost. Davison argued that it is hard to use Structure from Motion for real-time applications and proposed another method called Simultaneous Localisation and Mapping (SLAM) [1], which should be evaluated to be used for the sparse reconstruction.

Chapter 5

Conclusion

In this thesis, a method to do a variation based dense 3D reconstruction of a reference frame from a monocular mini-laparoscopic sequence has been implemented. It is able to cope with illumination changes by using photometric invariants. The reconstruction is done online and needs about 55 seconds for 30 frames. It has a median accuracy of 2.96 mm.

A reconstruction time of 55 seconds is far away from being real-time but it is much faster than the old implementation. It has been shown that more than 75% of the time is needed for the sparse reconstruction. It has to be considered whether the Structure from Motion approach is the right choice to be used for a real-time implementation. There is also the approach of Simultaneous Localization and Mapping (SLAM) which is better suited for real-time applications [1, 2, 23]. As the sparse reconstruction will always be needed for the initialization of the dense reconstruction, it has to be able to run in real-time. Besides the sparse reconstruction, the dense reconstruction can also not be considered being able to run in real-time at the moment. The longest step is the interpolation of an inverse depth map from the results of the sparse reconstruction. This step takes more than 6 seconds which is far too long. A nearest neighbour search is executed for each pixel on the CPU. This step should be transferred to the GPU in order to increase the performance. The other two steps of the dense reconstruction, update of the cost volume and minimization, take less than 3 seconds for the 30 frames. This is close to real-time which would be less than 2 seconds. In Section 4.4, it has been explicated that these steps have not been optimized so far and especially the performance of memory transfers could be increased.

During the performance evaluations, also an effect about the CUDA streams has shown up. Pushing one operation into a stream that contains no pending operations takes about 1 ms. Pushing an operation directly after another operation to the stream takes about 30 ms. This is a strange behaviour and can only be explained in a way that all preceding operations in a stream have to be finished to be able to push a new operation into the stream. This effect has to be examined and the working with streams has to be optimized.

The median accuracy of 2.96 mm is worse than the 0.89 mm that Marcinczak et al. got using their implementation [14], but there are improvements that can be made in the future. Section 4.2 has shown that the keyframe estimation is a problem in the sparse reconstruction. This problem caused many reconstructions to loose tracking, resulting in high errors. The keyframe estimation is not trivial for online reconstructions. There is no knowledge about the whole sequence or when the sequence will end. In offline reconstruction, it is possible to find an optimal path from the first to the last frame. In online reconstruction, only the current and some previous frames can be stored in the memory. Again, a look at SLAM might be useful to see how keyframe selection is done there or to replace the whole Structure from Motion approach with the SLAM approach.

The frame selection problem is also present in the dense reconstruction. The current implementation simply takes a number of frames directly following the reference frame. It is important that the frames used for computing the cost volume show the content of the reference frame from different viewports. Especially in the translational sequences, this is definitely not the case. A solution has to be found to better select the frames for computing the cost volume. It has to take into account the following and preceding frames, not only the following ones.

Figure 4.2 shows two other aspects that have to be improved. The first one is the coupling of the dense reconstruction to the sparse reconstruction. In the current evaluations, the coupling is too strong. It has to be examined whether it will be better to only initialize the inverse depth map with the interpolated one. The other aspect is the choice of data terms. Marcinczak et al. have shown that data terms based on the $r\phi\theta$ color space perform much better than data terms based on the RGB color space [14]. The problem with the $r\phi\theta$ color space is that similar dark colors can result in big differences in the angles ϕ and θ. Therefore, some other data terms have to be evaluated that are invariant to illumination changes, but also perform better in dark regions.

Besides these necessary steps to increase accuracy and performance, some extensions have to be done in order to reach the long term goal of being able to do quantitative measurements on the reconstructed surface. The first extension is to combine reconstructions of different reference frames into one surface. At the moment, only the content of one reference is reconstructed. Such a frame shows a small part of the liver. In order to get a reconstruction of the whole liver, it is necessary to do reconstructions on many frames and combine them to one surface. The second extension is to do tracking in the dense domain. Newcombe et al. have shown that a tracking in the dense domain is much more robust than a tracking in the sparse domain [18]. In that way, the sparse reconstruction is only executed at the beginning until a dense reconstruction is available and then the tracking is executed on that dense model. The last extension is scale estimation. Currently, the reconstruction is only estimated up to scale. There is no information on how big the reconstruction is or how far it is away from the camera. This extension is needed the most for quantitative measurements on the reconstructed surface but it is also the most complicated one.

Appendix A

Class Diagrams of CUDA Framework

exception
CudaException
-error: const string
+«constructor» CudaException(pError:const char*)
+«constructor» CudaException(pError:const cudaError_t&)
+what(): const char* const

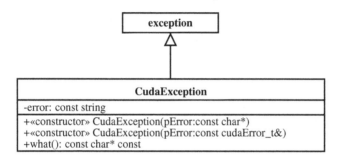

Figure A.1: Class diagram of the class *CudaException*.

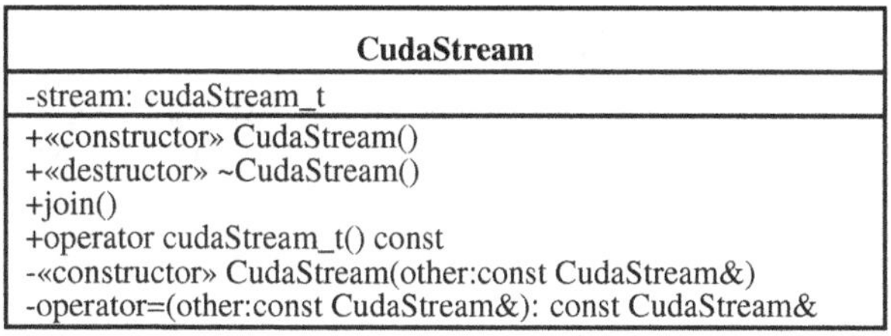

CudaStream
-stream: cudaStream_t
+«constructor» CudaStream()
+«destructor» ~CudaStream()
+join()
+operator cudaStream_t() const
-«constructor» CudaStream(other:const CudaStream&)
-operator=(other:const CudaStream&): const CudaStream&

Figure A.2: Class diagram of the class *CudaStream*.

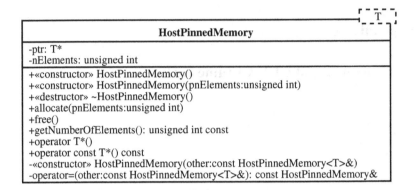

Figure A.3: Class diagram of the class *HostPinnedMemory*.

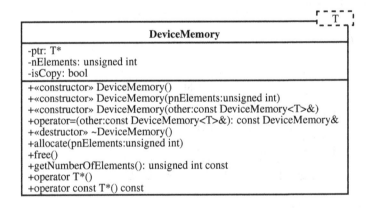

Figure A.4: Class diagram of the class *DeviceMemory*.

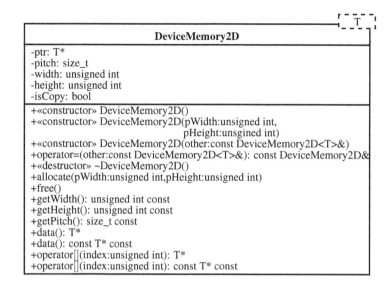

Figure A.5: Class diagram of the class *DeviceMemory2D*.

Appendix B

Class Diagrams of Sparse Reconstruction

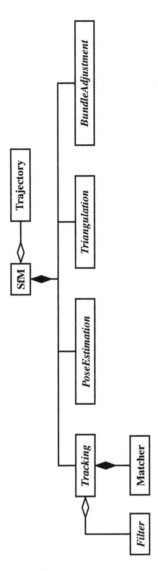

Figure B.1: Overview of the classes used in the sparse reconstruction. The reconstruction steps are implemented by using the strategy pattern. Only the base classes of the strategies are shown.

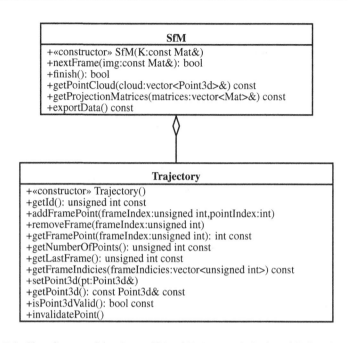

Figure B.2: Class diagram of the classes *SfM* and *Trajectory*. Only the public interface is shown.

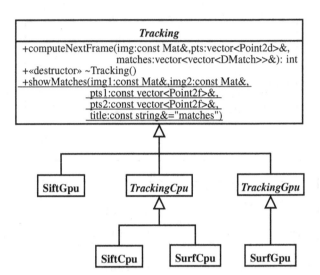

Figure B.3: Class diagram of the classes belonging to the tracking strategy. Only the public interface of the class *Tracking* is shown.

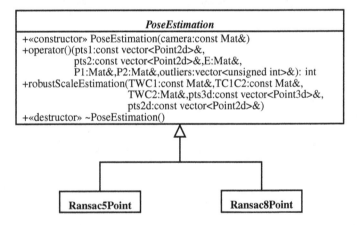

Figure B.4: Class diagram of the classes belonging to the pose estimation strategy. Only the public interface of the class *PoseEstimation* is shown.

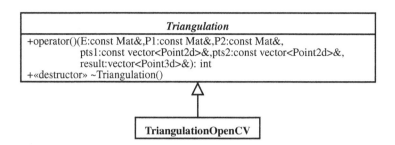

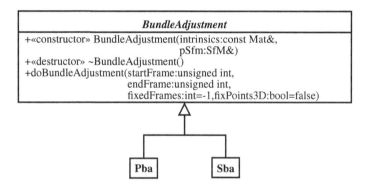

Figure B.5: Class diagram of the classes belonging to the triangulation strategy. Only the public interface of the class *Triangulation* is shown.

Figure B.6: Class diagram of the classes belonging to the bundle adjustment strategy. Only the public interface of the class *BundleAdjustment* is shown.

Appendix C

Evaluation Results of Sparse Reconstruction

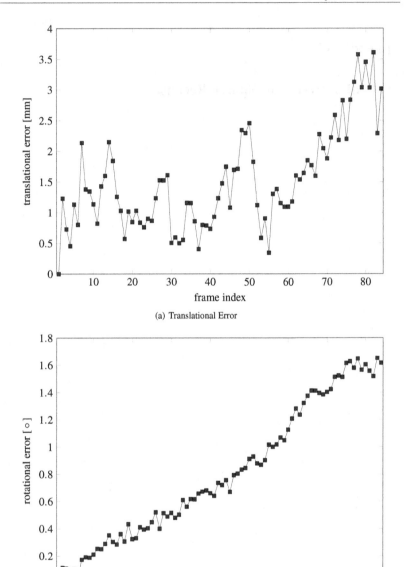

(a) Translational Error

(b) Rotational Error

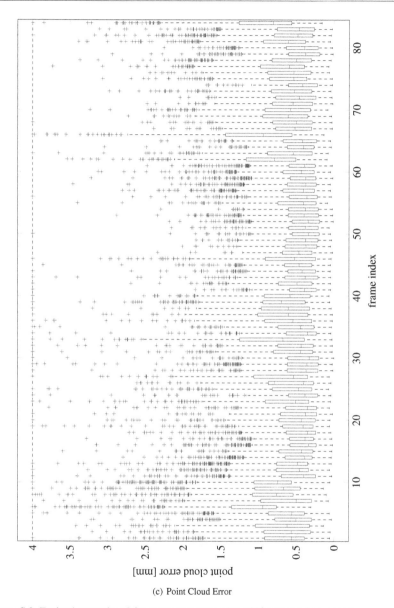

(c) Point Cloud Error

Figure C.1: Evaluation results of the sparse reconstruction of Phantom 1 Sequence 1.

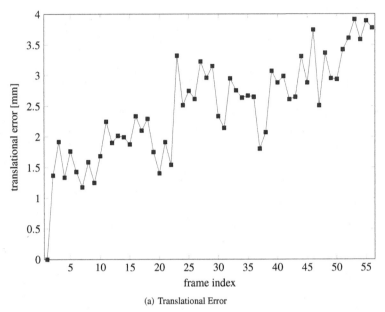

(a) Translational Error

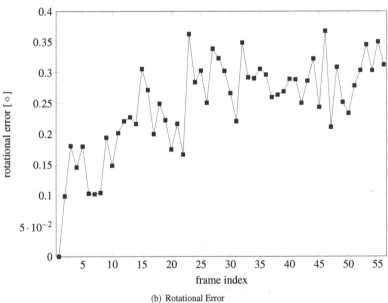

(b) Rotational Error

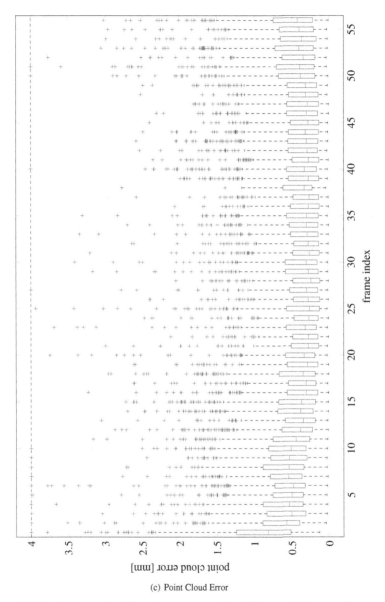

(c) Point Cloud Error

Figure C.2: Evaluation results of the sparse reconstruction of Phantom 1 Sequence 2.

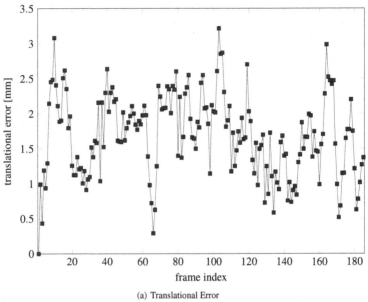

(a) Translational Error

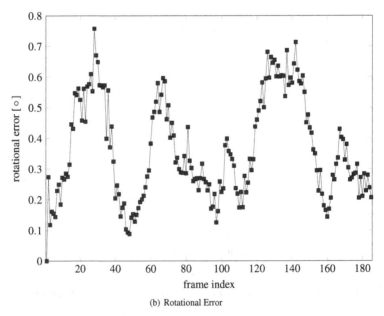

(b) Rotational Error

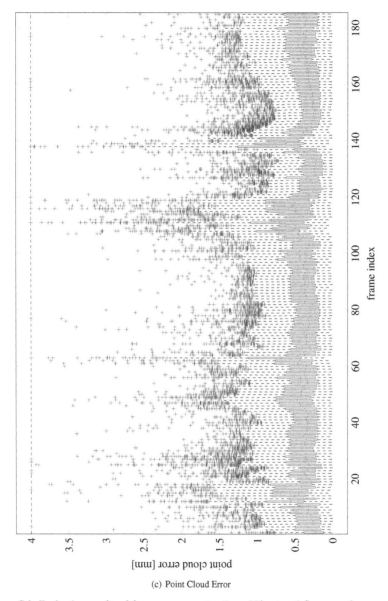

(c) Point Cloud Error

Figure C.3: Evaluation results of the sparse reconstruction of Phantom 1 Sequence 3.

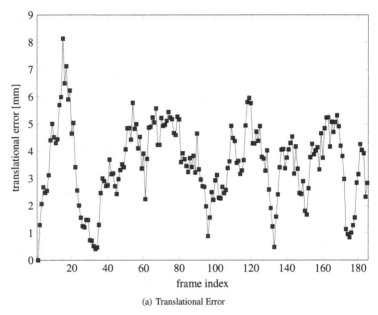

(a) Translational Error

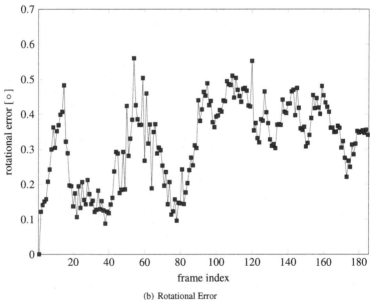

(b) Rotational Error

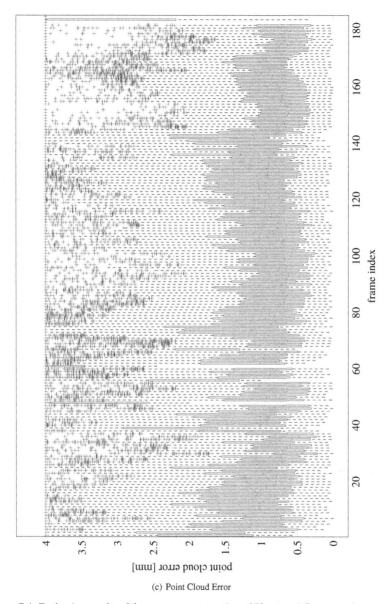

(c) Point Cloud Error

Figure C.4: Evaluation results of the sparse reconstruction of Phantom 1 Sequence 4.

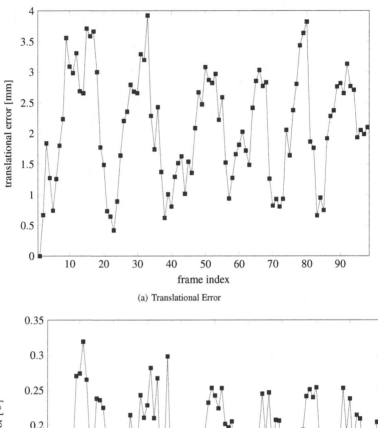

(a) Translational Error

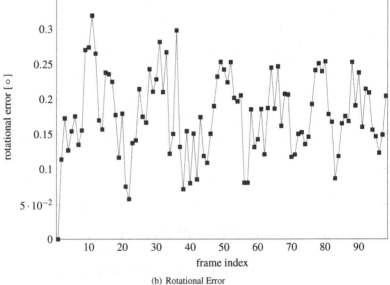

(b) Rotational Error

(c) Point Cloud Error

Figure C.5: Evaluation results of the sparse reconstruction of Phantom 1 Sequence 5.

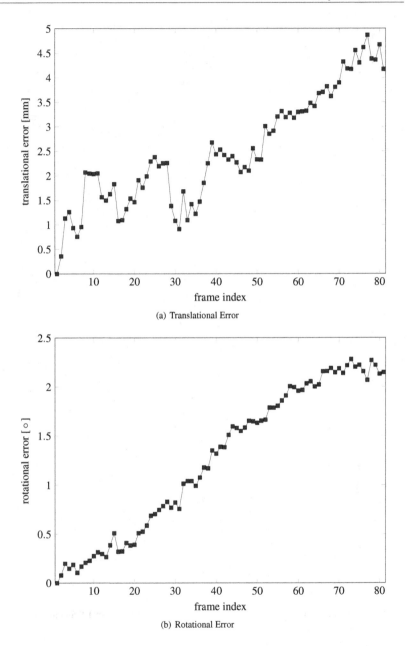

(a) Translational Error

(b) Rotational Error

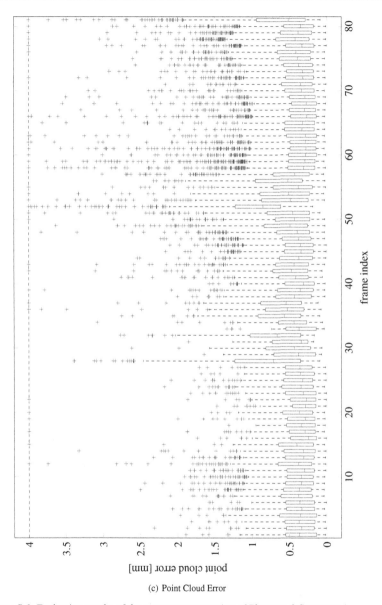

(c) Point Cloud Error

Figure C.6: Evaluation results of the sparse reconstruction of Phantom 2 Sequence 1.

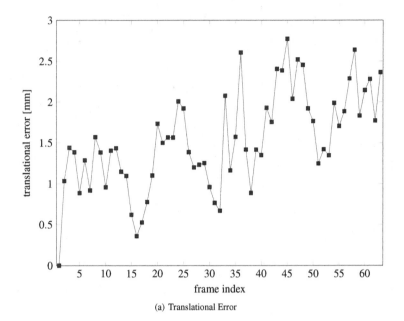

(a) Translational Error

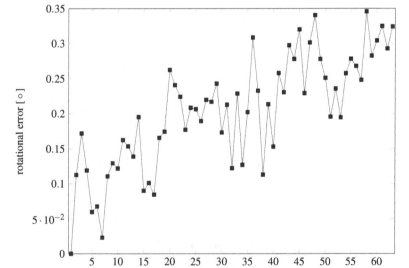

(b) Rotational Error

(c) Point Cloud Error

Figure C.7: Evaluation results of the sparse reconstruction of Phantom 2 Sequence 2.

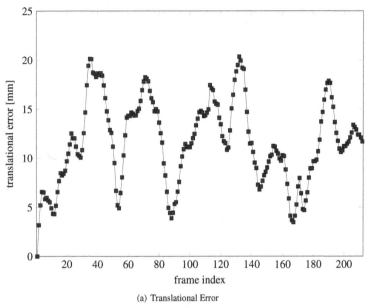

(a) Translational Error

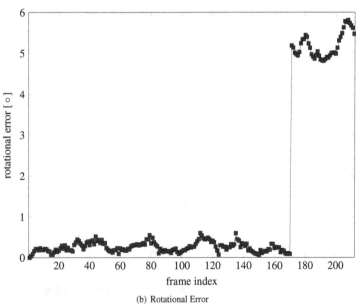

(b) Rotational Error

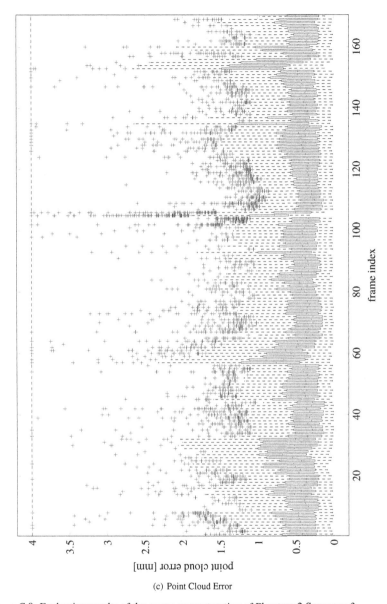

(c) Point Cloud Error

Figure C.8: Evaluation results of the sparse reconstruction of Phantom 2 Sequence 3.

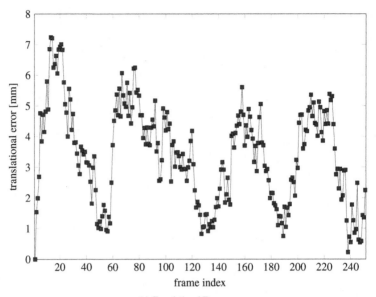

(a) Translational Error

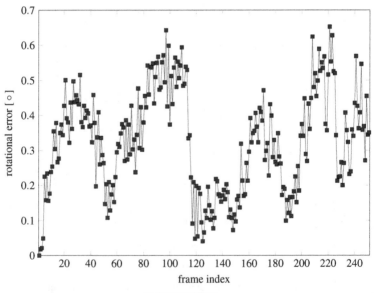

(b) Rotational Error

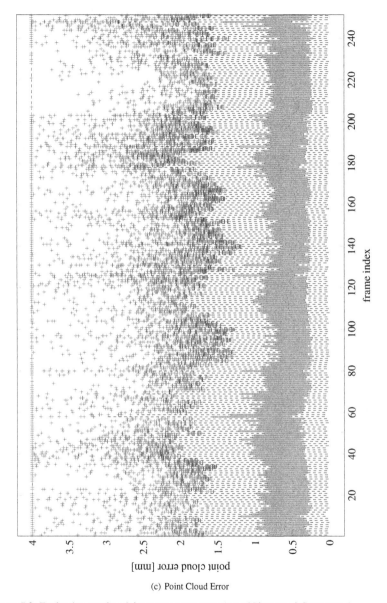

(c) Point Cloud Error

Figure C.9: Evaluation results of the sparse reconstruction of Phantom 2 Sequence 4.

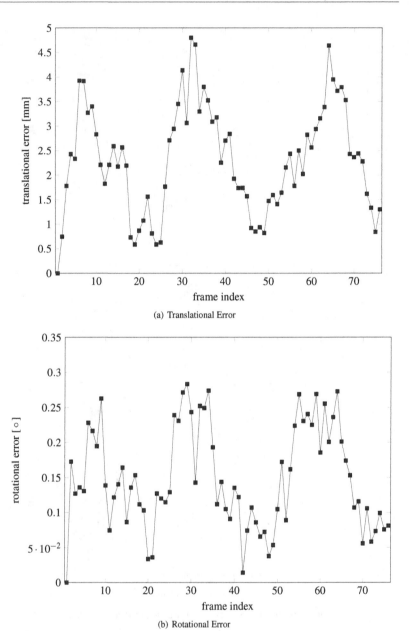

(a) Translational Error

(b) Rotational Error

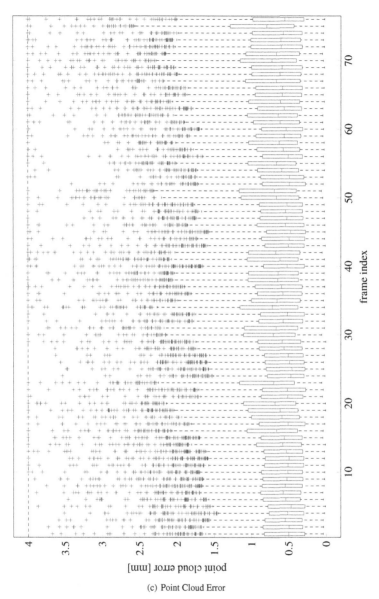

(c) Point Cloud Error

Figure C.10: Evaluation results of the sparse reconstruction of Phantom 2 Sequence 5.

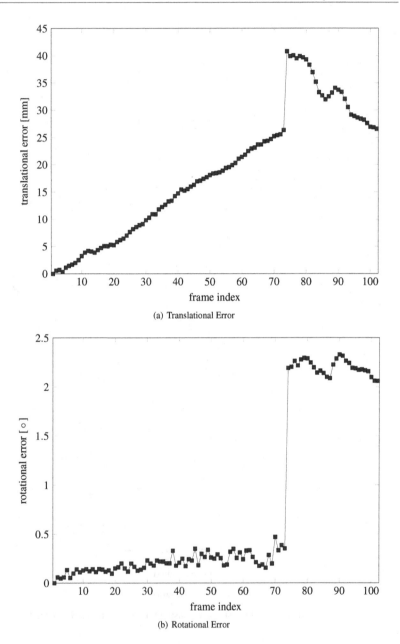

(a) Translational Error

(b) Rotational Error

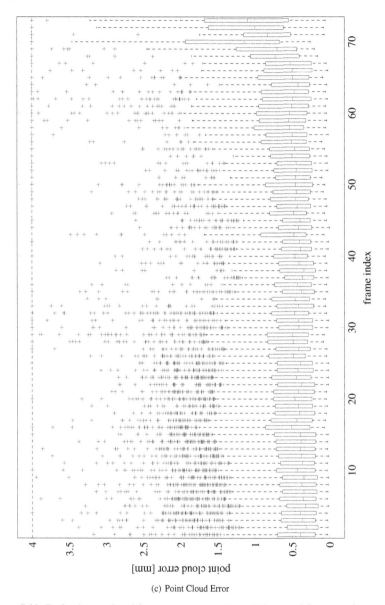

(c) Point Cloud Error

Figure C.11: Evaluation results of the sparse reconstruction of Phantom 3 Sequence 2.

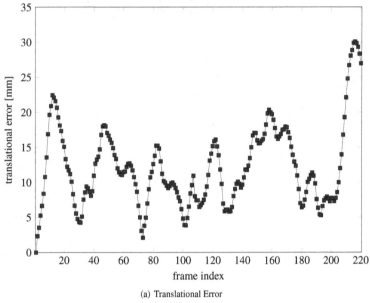

(a) Translational Error

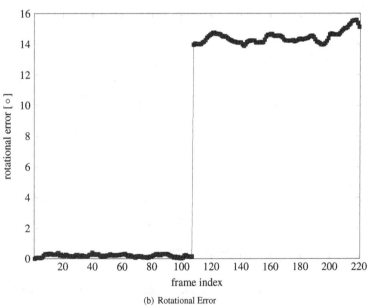

(b) Rotational Error

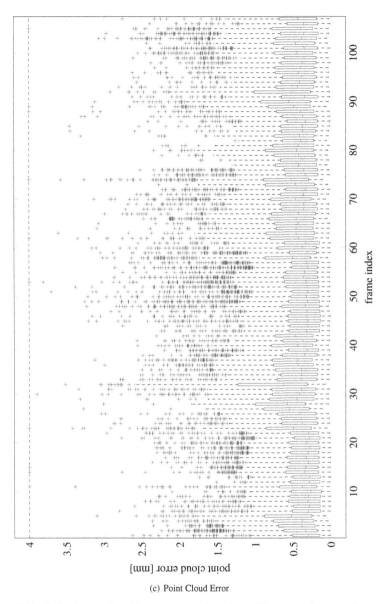

(c) Point Cloud Error

Figure C.12: Evaluation results of the sparse reconstruction of Phantom 3 Sequence 3.

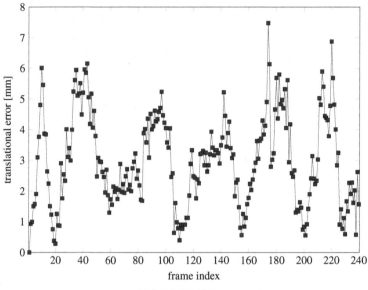

(a) Translational Error

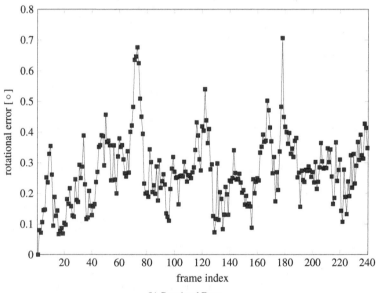

(b) Rotational Error

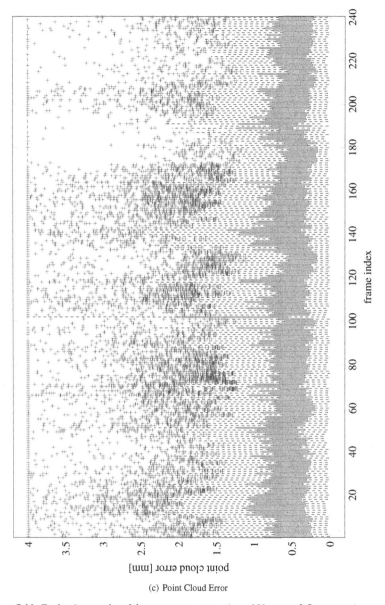

(c) Point Cloud Error

Figure C.13: Evaluation results of the sparse reconstruction of Phantom 3 Sequence 4.

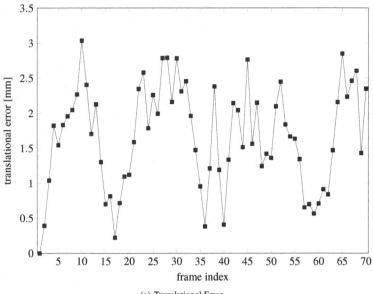

(a) Translational Error

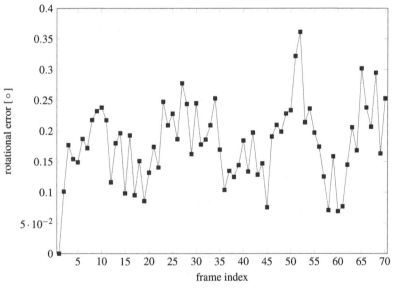

(b) Rotational Error

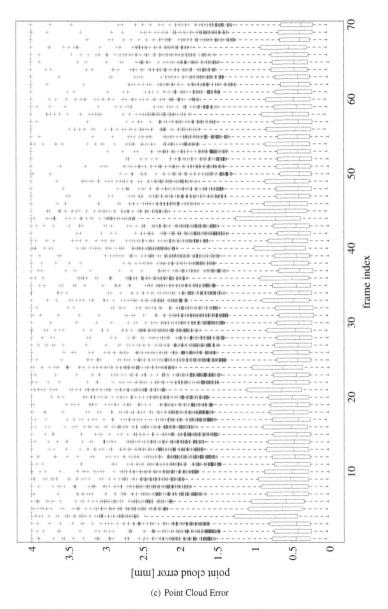

(c) Point Cloud Error

Figure C.14: Evaluation results of the sparse reconstruction of Phantom 3 Sequence 5.

Bibliography

[1] Andrew J. Davison. Real-Time Simultaneous Localisation and Mapping with a Single Camera. In *Proceedings of the 9th International Conference on Computer Vision, 2003*, pages 1403–1410. IEEE, 2003.

[2] Andrew J. Davison, Ian D. Reid, Nicholas D. Molton, and Olivier Stasse. MonoSLAM: Real-Time Single Camera SLAM. *IEEE Transactions on Pattern Analysis and Machine Intelligence*, 29(6):1052–1067, 2007.

[3] Paul Debevec, Tim Hawkins, Chris Tchou, Haarm-Pieter Duiker, Westley Sarokin, and Mark Sagar. Acquiring the Reflectance Field of a Human Face. In *Proceedings of the 27th annual conference on Computer graphics and interactive techniques*, pages 145–156. ACM Press/Addison-Wesley Publishing Co., 2000.

[4] Martin A. Fischler and Robert C. Bolles. Random Sample Consensus: A Paradigm for Model Fitting with Applications to Image Analysis and Automated Cartography. *Communications of the ACM*, 24(6):381–395, 1981.

[5] Erich Gamma, Richard Helm, Ralph Johnson, and John Vlissides. *Design Patterns: Elements of Reusable Object-Oriented Software*. Addison-Wesley, 1994.

[6] Ankur Handa, Richard A. Newcombe, Adrien Angeli, and Andrew J. Davison. Applications of Legendre-Fenchel transformation to computer vision problems. Technical report, Tech. Rep. DTR11-7, Department of Computing at Imperial College London, 2011.

[7] Richard Hartley and Andrew Zisserman. *Multiple View Geometry in Computer Vision*. Cambridge University Press, 2003.

[8] Berthold K. Horn and Brian G. Schunck. Determining optical flow. In *1981 Technical Symposium East*, pages 319–331. International Society for Optics and Photonics, 1981.

[9] Donna L. Hoyert and Jiaquan Xu. Deaths: Preliminary Data for 2011. *National Vital Statistics Reports*, 61(6), October 2012.

[10] Ansgar W. Lohse. Rolls Royce for everybody? Diagnosing liver disease by mini-laparoscopy. *Journal of Hepatology*, 54(3):584–585, 2011.

[11] M.I. A. Lourakis and A.A. Argyros. SBA: A Software Package for Generic Sparse Bundle Adjustment. *ACM Trans. Math. Software*, 36(1):1–30, 2009.

[12] David G. Lowe. Distinctive Image Features from Scale-Invariant Keypoints. *International Journal of Computer Vision*, 60(2):91–110, 2004.

[13] Jan Marek Marcinczak. *Image Analysis of Mini-Laparoscopic Sequences for Computer Aided Diagnosis of Liver Cirrhosis*. Verlag Dr. Hut München, 2015. ISBN 978-3-8439-2303-3.

[14] Jan Marek Marcinczak and Rolf-Rainer Grigat. Total Variation Based 3D Reconstruction from Monocular Laparoscopic Sequences. In *MICCAI 2014 Workshop on Abdominal Imaging: Computational and Clinical Applications*, volume 8676 of *Lecture Notes in Computer Science (LNCS)*. Springer, 2014.

[15] Isabel Cristina Patiño Mejía and Rolf-Rainer Grigat. Hand-Eye Calibration of a Laparoscope Using Optical Tracking Systems. Master thesis, Technische Universität Hamburg-Harburg, November 2013.

[16] Scott Meyers. *Effective C++: 55 Specific Ways to Improve Your Programs and Designs*. Addison-Wesley, 2005.

[17] Yana Mileva, Andrés Bruhn, and Joachim Weickert. Illumination-Robust Variational Optical Flow with Photometric Invariants. In *Pattern Recognition*, pages 152–162. Springer, 2007.

[18] Richard A. Newcombe, Steven J. Lovegrove, and Andrew J. Davison. DTAM: Dense Tracking and Mapping in Real-Time. In *IEEE International Conference on Computer Vision (ICCV), 2011*, pages 2320–2327. IEEE, 2011.

[19] David Nistér. An Efficient Solution to the Five-Point Relative Pose Problem. *IEEE Transactions on Pattern Analysis and Machine Intelligence*, 26(6):756–770, 2004.

[20] Lukas Petersen and Rolf-Rainer Grigat. Synchronisation der Bildakquisition von Endoskopen mit optischen Trackern. Bachelor thesis, Technische Universität Hamburg-Harburg, December 2012.

[21] Marc Pollefeys, Luc Van Gool, Maarten Vergauwen, Frank Verbiest, Kurt Cornelis, Jan Tops, and Reinhard Koch. Visual Modeling with a Hand-Held Camera. *International Journal of Computer Vision*, 59(3):207–232, 2004.

[22] Matthias Schlüter and Rolf-Rainer Grigat. Robuste Merkmalsverfolgung in laparoskopischen Videos. Bachelor thesis, Technische Universität Hamburg-Harburg, June 2013.

[23] Hauke Strasdat, José M. M. Montiel, and Andrew J. Davison. Visual SLAM: Why filter? *Image and Vision Computing*, 30(2):65–77, 2012.

[24] Udo von Öhsen and Rolf-Rainer Grigat. Ego-motion Estimation in Laparoscopic Videos. Master thesis, Technische Universität Hamburg-Harburg, June 2011.

[25] Andreas Wedel and Daniel Cremers. *Stereo Scene Flow for 3D Motion Analysis*. Springer, 2011.

[26] World Health Organization. Ten leading causes of death, 2012 (by sex). http://gamapserver.who.int/gho/interactive_charts/mbd/leading_cod/2012.asp. Last accessed on 30th October 2014.

[27] Changchang Wu. SiftGPU: A GPU Implementation of Scale Invariant Feature Transform (SIFT). http://cs.unc.edu/~ccwu/siftgpu, 2007. Last accessed on 2nd November 2014.

[28] Changchang Wu, Sameer Agarwal, Brian Curless, and Steven M. Seitz. Multicore Bundle Adjustment. In *IEEE Conference on Computer Vision and Pattern Recognition (CVPR), 2011*, pages 3057–3064. IEEE, 2011.

[29] Christopher Zach, Thomas Pock, and Horst Bischof. A Duality Based Approach for Realtime TV-L1 Optical Flow. In *Pattern Recognition*, pages 214–223. Springer, 2007.

Printed in the United States
By Bookmasters